W0106996

ULTRASONIC CUTTING

UL'TRAZVUKOVOE REZANIE

Ультразвуковое резание

ULTRASONIC CUTTING

by

**L. D. Rozenberg, V. F. Kazantsev,
L. O. Makarov, and D. F. Yakhimovich**

Authorized translation from the Russian
by J. E. S. Bradley, B.Sc., Ph.D.

*Preface to the English edition by
Lewis Balamuth, Ph.D.
Cavitron Ultrasonics*

Springer Science+Business Media, LLC
1964

The original Russian text, published for the Acoustics Institute by the Academy of Sciences Press in Moscow, in 1962, has been brought up to date by the authors for the English edition.

Л. Д. Розенберг, В. Ф. Казанцев,
Л. О. Макаров, Д. Ф. Яхимович
Ультразвуковое резание

Library of Congress Catalog Card Number 64-7762

ISBN 978-1-4899-4877-9   ISBN 978-1-4899-4875-5 (eBook)
DOI 10.1007/978-1-4899-4875-5

© *1964 Springer Science+Business Media New York*
Originally published by Consultants Bureau Enterprises, Inc in 1964.
Softcover reprint of the hardcover 1st edition 1964

All rights reserved

No part of this publication may be reproduced in any form without written permission from the publisher

PREFACE

"Ultrasonic Cutting" is a unique work in a number of respects. In the first place, there is no other work in English which is devoted wholly to a single specialized branch of ultrasonic technology. During the past ten years, numerous works have appeared on ultrasonics. Some emphasize pedagogical and theoretical aspects of the subject. Others are directed at engineers and the industrial public, so as to provide a bird's-eye view of the field. Such works, of course, do fill a real need – namely to acquaint the technical, industrial, and academic worlds with the broad development of a rapidly growing technology. But such works serve only as the barest introduction to a specific branch of ultrasonic practice such as ultrasonic machining, or impact grinding, as it is often called.

Also, when it is realized that practically no undergraduate or graduate university instruction is available in the field of ultrasonics, then any definitive work in an important area of this technology takes on an added importance. The authors of "Ultrasonic Cutting" have looked deeply and intensively into the process of ultrasonic machining, and they have kept abreast of developments in other countries, as is witnessed by the extensive bibliographies accompanying each chapter of the book.

The result is a unique work which deserves careful study by the reader. Attention is given to the elementary theories needed to grasp the basic scientific character of ultrasonics as used in machining. At the same time, most of the important practical questions which one working in the field would want to ask are raised and are illuminated with quantitative findings and "hardware" designs.

The book presents a logical development of its subject, starting in the first chapter with basic information covering definitions, concepts, and equations needed to understand the high-frequency energy transformations encountered in ultrasonic impact grinding. The elementary theory of the vibrations of a simple one-degree-of-freedom system is covered, as well as the simple theory of wave propagation in liquids and gases. The ideas are extended to simple solid waveguides, such as a rod whose diameter is smaller than a half-wavelength of compression waves in the material. Finally, the character of cavitation is briefly considered as it arises in a liquid due to high-intensity waves therein. The selection of material in this chapter is adequate for the practical understanding of impact grinding.

The second chapter presents original work initiated to elucidate the nature of ultrasonic impact grinding. It also includes the work of others which is covered in the text as well as in the appended bibliography. The picture developed here quite conclusively shows impact grinding to be due primarily to the repeated successive blows of the loose abrasive grains on the work surface as these grains are hammered by the tool surface. Cavitation is shown to have an important but secondary role in the process. The direct visual evidence of high-speed cinematography is illustrated and lays the basis for future theorizing. The theorizing presented shows reasonably good accord with experimental results.

The remaining three chapters contain the practical, engineering heart of the book. They present in order the theory and design of the acoustical section of an ultrasonic machine tool, the design and description of ultrasonic machining tools, and finally the technology of ultrasonic machining.

The coverage of material is such that the book provides a good reference source for current problems in impact grinding. The virtues and shortcomings of the method are frankly and clearly faced. There is brief consideration given to other new methods of material removal such as electrospark and electrochemical methods of processing. The combination of these electrical methods with ultrasonic impact grinding is mentioned.

The material is ample for those who desire to enter this field and use it in a practical way. Of course, the size of the work precludes any attempt at complete coverage of the art. Nevertheless, the broad guidelines of possible future development are there. All in all, the virtues of "Ultrasonic Cutting" far outweigh such weaknesses as insufficient treatment of some cutting problems and a too cursory description of the electronic generator equipment needed to generate the vibration. The authors clearly are well grounded from personal experience in the design and use of the acoustical and mechanical parts of an ultrasonic machine tool. I doubt that their experience is so thorough in the electronic design aspects. This is a common situation to be found wherever impact grinding is practiced because the electronic equipment is often obtained by purchase or by assigning its development to an electronic engineering group not directly concerned with the machining problems themselves.

L. Balamuth

FOREWORD TO THE ENGLISH EDITION

Although not too much time has elapsed since the publication of the Russian edition, we have felt it desirable to make a few small alterations in order to give an adequate picture of recent progress in ultrasonic cutting. Some misprints and minor errors have also been corrected.

It is hoped that the American reader will find much of value and interest in it not only as regards developments in this country but also in others.

The Authors

FOREWORD

Current plans call for a vast expansion of the national economy of the USSR; in particular, machine-construction industries must design and produce the necessary machines, which must utilize the latest advances in science and technology, including ultrasonics.

One of the most interesting and valuable industrial applications of ultrasonics is the process known as ultrasonic cutting or ultrasonic dimensional working of brittle materials. This process is used here and elsewhere; standard machines are available, and several hundred papers have been published on various aspects of the process. All the same, the process cannot be said to have been thoroughly studied and completely worked out; the results and suggestions of the various workers are often in conflict.

There is an urgent need to introduce the process widely, but the material to be surveyed is extensive, and the two together make it essential to adopt a unified point of view, because the large volume of material already in existence acts as a brake on further work and also prevents wide circles of scientists and technologists from becoming fully acquainted with this interesting process.

Here we attempt to survey and review all the available information on the various aspects of the process. The book is based on the experience of the authors in the Ultrasonics Laboratory at the Acoustics Institute, Academy of Sciences of the USSR, and in the Special Designs Office of the Moscow City Economic Council, but we have tried to incorporate also all relevant work in this field done in the USSR and elsewhere.

The book is directed to scientific workers, design engineers, and graduate students. The first chapter is directed to those who are not specialists in this field; it presents a general survey of mechanical vibrations and waves, which is essential to a proper understanding of the main body of the text.

CONTENTS

INTRODUCTION

Current techniques for mechanical working of ordinary materials are highly developed, and machine tools have been greatly improved in recent years; one is now able to solve many of the varied and complex problems raised by rapid advances in technology.

However, recent years have seen the introduction of many new materials, many of which are very difficult to work. This applies particularly to superhard materials, such as tungsten and titanium carbides, diamonds, rubies, sapphire, hard steels, magnetic alloys, and corundum. Parts made of these materials can be worked only by grinding.

Another group of materials (germanium, silicon, ferrites, ceramics, glass, quartz) gives difficulty on account of great brittleness; these materials often cannot withstand the forces needed for mechanical working.

The need for methods of working these "unworkable" materials has led to the introduction of special methods (electrochemical, electroerosion, electron-beam, and so on) [1]. Ultrasonic cutting is one of these.

The ultrasonic method is a form of abrasion; the brittle material is removed by blows from grains of a harder abrasive, which is under the control of a tool which vibrates with a comparatively small amplitude. Of course, the abrasive also causes wear in the tool, but this is minimized by making the tool of a viscous material (one that has no tendency to cleave). The particles of abrasive are themselves cleaved in the process and so must be gradually replaced by running into the working area a liquid carrying fresh abrasive, which also serves to flush away the products. The material is cut away as very small particles, but these are produced by many abrasive grains, and the tool vibrates at a high frequency, so the total rate of removal can be sufficient for practical purposes.

The tool may be advanced in the direction of vibration, in which case there is produced a cavity whose profile corresponds precisely to that of the tool. Combinations of movements allow one to perform a variety of operations on brittle materials analogous to those of ordinary milling, shaping, profile milling, and so on.

The noise resulting from this process is minimized by choosing a frequency in the low ultrasonic range (16-25 kc), so the method is one of ultrasonic working.

The method in the form described above was proposed by Farrer* in 1945 [2].

---

*Rozenberg's crediting to Farrer is an exquisite example of the unfortunate high frequency of "noise" when scientific information crosses language barriers. Farrer was the patent agent on the first issued patent, British patent No. 602,801 (1945), issued to L. Balamuth, who discovered ultrasonic machining accidentally in 1942, while he was investigating the dispersion of solids in liquids by means of a magnetostrictively vibrating nickel tube. The United States Patent for the process, No. 2,580,716, was issued in 1962.

Cavitron Ultrasonics, Inc., is the present name of the company formed to develop the practical use of this invention. The Sheffield Corporation and the Raytheon Mfg. Co. are exclusive licensees of Cavitron and are the only licensed manufacturers of ultrasonic impact grinding equipment in the United States. At present the process is in use in this country in dental drilling and dental prophylaxis, machining die cavities in hard metals, machining ceramics and other nonconductive materials, dicing semiconductor blanks for solid state devices, machining cemented carbides for special nozzles and other parts, and, of course, cutting glass and gem stones. – Publisher.

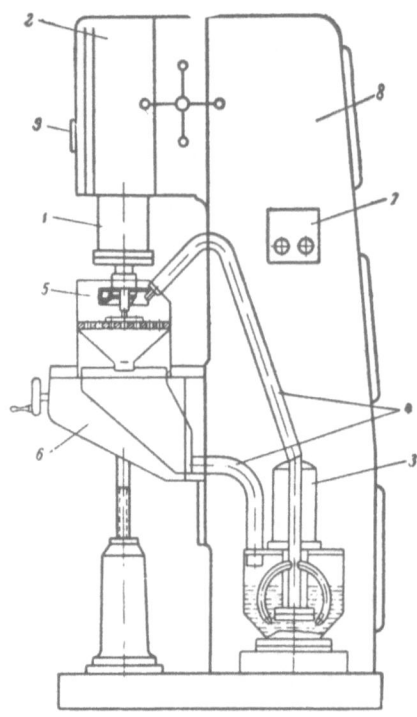

Fig. 1. Current machine
for ultrasonic cutting.

The valuable features of the method (possibility of working superhard and very brittle materials, ease of piercing and recessing, precision working) immediately attracted attention; the industrial use of ultrasonic cutting has advanced rapidly.

The first communications on equipment and techniques for ultrasonic cutting appeared in 1951-1952; by 1953-1954 the first ultrasonic machine tools had been made, mostly on the basis of drilling and milling machines. This was the period of the early development of ultrasonic cutting; recent years have seen the introduction of ultrasonic machine tools of various types and sizes for a variety of purposes. Some models have begun to come into regular production, and detailed studies have begun on the physics of ultrasonic cutting; much experience on the design of ultrasonic machines has accumulated.

The rapid progress in this field is clear from the number of published papers. In 1954 Bergmann [3] listed only nine on ultrasonic machining, whereas now there are over 300, apart from patents.

Many advances in our understanding of the process and in the design of equipment have been made in this country; several institutes and design organizations are now concerned with research on the process and development of equipment and methods.

Research on ultrasonic cutting methods and equipment is also proceeding in the countries of the people's democracy.

The principal producers of equipment in capitalist countries are Sheffield and Raytheon in the USA, Mullard and Kerry in Great Britain, Lehfeldt in West Germany, and certain firms in France and Italy. Some interesting research on the process has also been done in Japan.

Notable treatments of the physics of the process have been given by Nomoto [4], Neppiras [5], Goetze [6], Dikushkin and Barke [7], Shaw [8], and Miller [9]. The process as a whole is complex, and there are several hypotheses as to the physics of the processes. Rozenberg and Kazantsev [10] have used high-speed cinematography to elucidate this; D'yachenko et al. [11] have done similar work.

Other high-frequency machining methods have been examined; in 1948 Rosenthal [12] patented a machine for machining without the use of abrasive by means of a tool vibrating with a frequency of 30,000 to 1,000,000 c/s. Abrasive was used only for finishing and lapping. The source was a magnetostriction or piezoelectric vibrator.

The use of an abrasive paste activated by ultrasonics has also been described [13], but this has not found industrial use.

Attempts have been made to accelerate the process by combining the ultrasonic vibrations with low-frequency ones and with electrical pulses [14], with electrochemical machining [15], and with electrospark methods [16-19].

Current machines (Fig. 1) have several specific features: an acoustic head 1, feed mechanism 2, abrasive feed system (including pump 3, pipes 4, and jet 5), and power source (not shown). In addition, there are features found in ordinary machine tools, such as the table 6, control panel 7, frame 8, and position indicator 9.

The acoustic head contains the electromechanical converter, which drives the tool via a special holder (waveguide). The feed mechanism applies the necessary force (5-8 kg) between tool and workpiece. The abrasive feed system continuously brings in fresh abrasive to the cutting area, removes products, and cools the components.

The power source supplies the ultrasonic current to the acoustic head; various vacuum-tube systems are in common use, and some high-power magnetostriction heads have recently been used with high-frequency alternators, which appear very promising.

Ultrasonic cutting is a technique as yet far from perfected, and existing machines have many deficiencies: they are not very reliable, are costly, and are of very low efficiency. Ultrasonic cutting techniques are only beginning to be exploited. No really reliable methods are available for calculating the dimensions of components, especially cutting tools.

The method has many valuable features, which ensure it a place among specialized machining methods; it is bound to develop in the near future, for it enables one to perform operations that cannot be performed by any other method.

Literature Cited

1. New Developments in Electrical and Ultrasonic Machining of Materials, Collection, L. Ya. Popilov (ed.), Lenizdat, 1959.
2. J. O. Farrer, Method of Abrading, British Patent No. 602801, 1948 (submitted April 14, 1945).
3. L. Bergmann, Ultrasonics [Russian translation], IL, 1956.
4. A. Nomoto, "Ultrasonic machining by low power vibration," J. Acoust. Soc. Am. 26(6):1081, 1954.
5. E. A. Neppiras, "Report on ultrasonic machining," Metalwork. Product. 100(27-31):33, 34, 1956.
6. D. Goetze, "Effect of vibration amplitude, frequency, and composition of abrasive slurry on the rate of ultrasonic machining in ketos tool steel," J. Acoust. Soc. Am. 28(6):1033, 1956.
7. V. I. Dikushin and V. N. Darke, "Ultrasonic erosion and its relation to vibrational parameters of the tool," Stanki i Instr., No. 5:10, 1958.
8. M. C. Shaw, "Das Schleifen mit Ultraschall," Microtechnic 10(6):265, 1956.
9. G. E. Miller, "Special theory of ultrasonic machining," J. Appl. Phys. 28(2):149, 1957.
10. L. D. Rozenberg and V. F. Kazantsev, "Physics of the ultrasonic machining of brittle materials," Doklady Akad. Nauk SSSR 124(1):79, 1959.
11. V. G. Aver'yanov and A. A. Milovidov, "A study of ultrasonic machining processes," in: Summaries of papers at the Third Conference on High-Speed Photography and Cinematography, Moscow, Izd. Akad. Nauk SSSR, 1960.
12. A. H. Rosenthal, Machine for mechanically working materials, US Patent No. 2452211, dated October 26, 1948.
13. I. B. Aronov and N. I. Siyak, Apparatus for ultrasonic grinding and polishing, Author's certificate, USSR No. 109101, dated November 29, 1956.
14. Methods of Accelerating Ultrasonic Machining, Moscow, Izd. TsBTI ÉNIMS, 1959.
15. A. I. Markov and B. N. Lyamin, Method of ultrasonic machining, Authors' certificate, USSR No. 209844, dated January 14, 1957.
16. G. Nishimura, K. Yanagishima, and T. Shima, "Ultrasonic electro-spark machining," J. Fac. Eng., Univ. Tokyo 25(1):41-46, 1957; 25(4):237, 253, 1958.
17. G. Nishimura and S. Shimakawa, "Ultrasonic electro-spark machining," J. Fac. Eng., Univ. Tokyo 25(4): 243, 247, 1958.
18. M. P. Higgins and C. E. Comstock, Machine tool, US Patent No. 2766364, dated November 8, 1952.
19. P. Gagnaire, "Les ultrasons dans le cadre industriel," Ind. franc. Achats et entret. matér. industr. 8(87): 507, 509, 511, 1959.

BASIC INFORMATION
ON MECHANICAL VIBRATIONS AND WAVES

The cutting process employs the mechanical vibrations of the tool; the process cannot be of high efficiency unless most of the mechanical energy in the vibrator is transmitted to the working end of the tool, so losses in the vibrator itself and via the holders must be minimized. The efficiency of the vibrator itself is dependent on the load to which it is coupled; best efficiency is ensured by matching the two.

Chapter 3 deals with the oscillatory systems, which are complicated and composite. Here we consider the main features of vibrations and elastic waves, as well as some aspects of the propagation of high-intensity waves in liquids.

1. Vibrations of a Simple System

An elastically restrained mass (Fig. 2) is a very simple mechanical oscillatory system. (It is assumed that the motion is horizontal; gravity can then be neglected.) If the ball is displaced and then released, it will oscillate in the horizontal plane about its equilibrium position; these oscillations would persist for an infinitely long time if there were no energy losses. These are undamped oscillations. The displacement ξ at a given time is

$$\xi = \xi_m \cos \omega_0 t, \tag{1.1}$$

in which $\omega_0 = 2\pi f_0$ is the angular frequency and ξ_m is the amplitude (initial displacement); f_0 is the frequency (number of complete oscillations in 1 sec).*

Figure 3 shows the waveform, with deviation from the equilibrium position (plotted vertically) against time (horizontally).

The differential equation for the motion is readily derived by equating the sum of the internal forces to zero; the external forces are zero, except at the instant of the original displacement.

In this case we have an inertial force F_u and an elastic force F_χ, so that

$$F_u + F_\chi = 0. \tag{1.2}$$

Newton's second law gives

$$F_u = M \frac{d^2\xi}{dt^2}, \tag{1.3}$$

in which M is the mass of the sphere and $d^2\xi/dt^2$ is the acceleration.

Hooke's law gives the elastic force as

$$F_\chi = \chi\xi, \tag{1.4}$$

in which ξ is displacement and χ is coefficient of elasticity, being the ratio of the elastic force to the deformation producing it. This coefficient is in dyne/cm in the cgs system, and in kg/m in the MKS system.

From (1.3) and (1.4) we have the differential equation as

$$M \frac{d^2\xi}{dt^2} + \chi\xi = 0. \tag{1.5}$$

---

* The second has been taken as the basis for the unit of frequency (c/s); 1000 c/s is abbreviated as 1 kc.

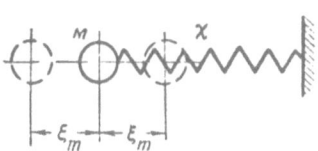

Fig. 2. A simple oscillatory system.

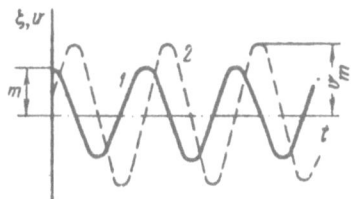

Fig. 3. Vibrations of a simple system: 1) variation in displacement; 2) variation in speed.

This is best reduced to canonical form (by dividing all terms by M) for integration

$$\frac{d^2\xi}{dt^2} + \frac{\chi}{M}\xi = 0.$$

This χ/M is a constant usually denoted by ω_0^2, so*

$$\frac{d^2\xi}{dt^2} + \omega_0^2\xi = 0. \qquad (1.5a)$$

(1.5a) is an ordinary linear differential equation of second order and is characteristic of undamped harmonic vibrations. The conditions at the start are t = 0, $\xi = \xi_m$; in this case (1.1) is the integral of (1.5a).

The physical significance of ω_0 is that of the angular frequency of the natural vibration of the system; it is not dependent on the initial displacement:

$$\omega_0 = \sqrt{\frac{\chi}{M}} \quad \text{or} \quad f_0 = \frac{1}{2\pi}\sqrt{\frac{\chi}{M}}. \qquad (1.6)$$

Here χ is expressed in dyne/cm and M in g; the velocity is the first derivative of the displacement with respect to time:

$$v = \frac{d\xi}{dt} = -\omega_0\xi_m \sin \omega_0 t = -v_m \sin \omega_0 t. \qquad (1.7)$$

This velocity is also periodic; it is largest as the ball passes through its equilibrium position, and it is zero (changes sign) at the maximum displacement. The largest value of the velocity is called the amplitude of the velocity v_m, which is proportional to the frequency and the amplitude of the displacement if the motion is harmonic: $v_m = \omega_0 \xi_m$.

The acceleration is the first derivative of the velocity with respect to time:

$$u = \frac{dv}{dt} = -\omega_0 v_m \cos \omega_0 t = -u_m \cos \omega_0 t. \qquad (1.8)$$

This is also periodic; its amplitude $u_m = \omega_0^2 \xi_m$ is proportional to the square of the frequency and to the first power of the amplitude of the displacement.

A real system always has losses, such as friction with the external medium, internal friction in the spring, and effects of any load imposed. The oscillations then decay, for the energy is consumed in overcoming friction or in doing external work; each maximal displacement is then less than the previous one (Fig. 4).

The differential equation for this case is readily found by introducing a frictional force, which in many cases is proportional to the velocity:

$$F_r = r\frac{d\xi}{dt}. \qquad (1.9)$$

---

* The square emphasizes that χ/M is always positive.

5

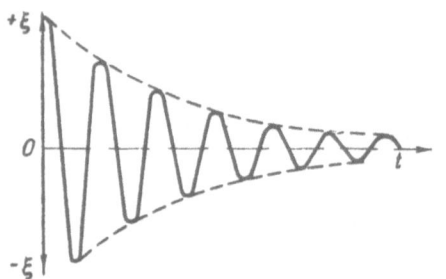

Fig. 4. Damped oscillations. The dotted line denotes the envelope, whose form is $e^{-\delta t}$.

With $\chi /M = \omega_0^2$, we have

Here $d\xi/dt = v$ is the velocity and r is a coefficient of proportionality (frictional coefficient), whose dimensions are dyne-sec/cm and kg-sec/m in the cgs and MKS systems, respectively.

The equation for equilibrium is

$$F_u + F_\chi + F_r = 0,$$

or, with (1.3), (1.4), and (1.9),

$$M \frac{d^2\xi}{dt^2} + r \frac{d\xi}{dt} + \chi\xi = 0. \qquad (1.10)$$

In canonical form

$$\frac{d^2\xi}{dt^2} + \frac{r}{M} \frac{d\xi}{dt} + \frac{\chi}{M} \xi = 0.$$

$$\frac{r}{M} = 2\delta, \qquad (1.11)$$

and so finally

$$\frac{d^2\xi}{dt^2} + 2\delta \frac{d\xi}{dt} + \omega_0^2\xi = 0. \qquad (1.10a)$$

This δ is proportional to r and is called the logarithmic decrement.

Then, with $\xi = \xi_m$ for t = 0 we have the integral of (1.10a) as

$$\xi = \xi_m e^{-\delta t} \cos\sqrt{\omega_0^2 - \delta^2}\, t. \qquad (1.10b)$$

This is the analytic form of the graph in Fig. 4; δ defines the slope of the envelope and indicates the loss in amplitude per cycle. Its relation to r is given numerically by

$$\delta = \frac{r}{2M}. \qquad (1.11a)$$

The larger the loss the larger the decrement, so the more rapidly the oscillations die away.

Another important parameter is what is called the system magnification Q:

$$Q = \frac{\omega_0 M}{r} = \frac{\omega_0}{2\delta}. \qquad (1.12)$$

This represents the ratio of the stored energy to the rate of loss of energy from all causes (including external work). Its value is infinitely large for a system free from losses.

The frequency of vibration in a system with losses is

$$f_n = \sqrt{f_0^2 - \left(\frac{\delta}{2\pi}\right)^2}. \qquad (1.13)$$

The losses thus reduce the frequency as well as damp the oscillations.

The second term under the root can be neglected if the loss is small, in which case $f_n = f_0$.

A system with losses can be kept oscillating by supplying it with energy to balance the loss; for this we need an external sign-varying (harmonic) force, whose instantaneous value may be put as

$$F = F_m \cos\omega_e t, \qquad (1.14)$$

in which F_m is amplitude and ω_e is angular frequency.

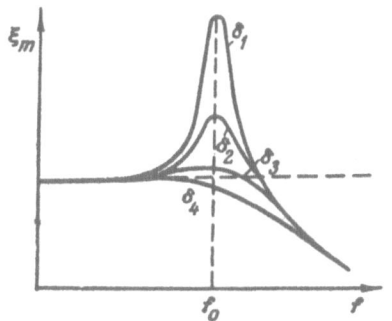

Fig. 5. Resonance curves for the amplitude of displacement for various degrees of damping ($\delta_1 < \delta_2 < \delta_3 < \delta_4$).

The oscillations this produces are called forced, and their frequency is that of the external force. [*]

The differential equation for this case is readily derived from (1.10) by equating the internal forces in the system not to zero but to the external force F.

From (1.10) and (1.14)

$$M \frac{d^2\xi}{dt^2} + r \frac{d\xi}{dt} + \chi\xi = F_m \cos \omega_e t \tag{1.15}$$

or, after reduction to canonical form as above,

$$\frac{d^2\xi}{dt^2} + 2\delta \frac{d\xi}{dt} + \omega_0\xi = \frac{F_m}{M} \cos \omega_e t. \tag{1.15a}$$

The steady-state solution is

$$\xi = \xi_e \cos (\omega_e t + \psi), \tag{1.16}$$

which shows that the oscillations are undamped and have the frequency ω_e of the external force, being shifted in phase by ψ relative to that force. This phase shift is usually without significance, so we omit it in what follows.

The amplitude ξ_e is

$$\xi_e = \frac{F_m}{4\pi^2 M \sqrt{\left(f_0^2 - f_e^2\right)^2 + \left(f_e \frac{\delta}{\pi}\right)^2}} . \tag{1.17}$$

Figure 5 shows a family of curves for the amplitude as a function of f_e for several values of the decrement. All curves run as one parallel to the x axis for frequencies well below the natural frequency, so here the amplitude is not dependent on the natural frequency. In fact, if in (1.17) we neglect all terms in the denominator except that in f_0^2 and give this its value from (1.6), we have for ξ_e in this range:

$$\xi = \frac{F_m}{\chi} . \tag{1.18}$$

This is simply the static deflection that a force F_m would produce in a spring with an elastic coefficient χ

The amplitude of the oscillations rises rapidly to a peak at $f_e = f_0$.

The increase in amplitude near the natural frequency is termed mechanical resonance. The amplitude at resonance is given by (1.17) with $f_e = f_0 = f_r$:

$$\xi_e = \frac{F_m}{4\pi f_r \ \delta M} = \frac{F_m}{\omega_r \ r} . \tag{1.19}$$

An ideal system (r = 0) would have an infinitely great amplitude, and so the losses must be minimized if we wish to obtain the largest possible amplitude from a real system; a system with large losses gives a lower amplitude at resonance. If the loss is very high (bottom curve in Fig. 5), there is no resonance at all; there is merely a slow fall in response towards high frequencies.

The amplitude is inversely proportional to the square of the frequency above the resonance:

$$\xi_e = \frac{F_m}{4\pi^2 f_e^2 M} = \frac{F_m}{\omega_e^2 M} . \tag{1.20}$$

Then the amplitude is governed by each one of the three parameters in turn in the three frequency ranges: by χ at low frequencies, by r near resonance, and by M above resonance.

---

[*] Oscillations of frequency ω_0 arise at the instant when the external force is applied, but these soon die away at a rate governed by δ.

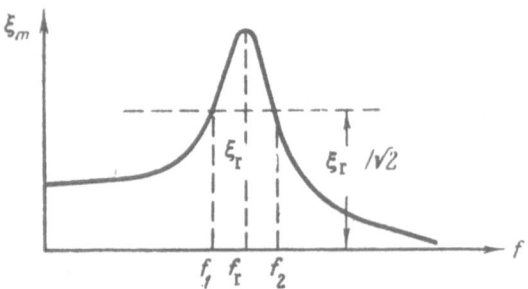

Fig. 6. Deduction of magnification
from the resonance curve.

The width of the peak is governed by the decrement or magnification, so these can be deduced from the curve, for which purpose an intercept is drawn at a height $\xi_r/\sqrt{2}$ (Fig. 6). Then we have

$$Q = \frac{f_r}{f_1 - f_2} = \frac{\omega_r}{\omega_1 - \omega_2}. \tag{1.21}$$

The behavior is thus governed by the frequency and by the system parameters, and, in particular, by the relation between the external frequency and the natural frequency of the system.

Eq. (1.17) shows that the amplitude of the motion is always proportional to that of the force; the ratio of these is a function solely of the system parameters and frequency, and it is not dependent on the magnitude of the force. From (1.6) and (1.11a) we can rewrite (1.17) as

$$\frac{F_m}{\xi_e} = \sqrt{(\chi - \omega_e^2 M)^2 + (\omega_e r)^2}. \tag{1.22}$$

The left-hand side in (1.22) is sometimes called the acoustic rigidity, by analogy with the static rigidity, which is the ratio of the force to the deformation it produces. The dimensions are dyne/cm or kg/m, as for the coefficient of elasticity.

The amplitude produced in a system having a given frequency by a fixed force is inversely related to the acoustic rigidity. Figure 7 shows the acoustic rigidity as a function of frequency for several values of r; this is simply the reciprocal of the curves in Fig. 5 for the amplitude as a function of frequency.

A more convenient characteristic is the mechanical impedance (or resistance), which is the ratio of the external force to the velocity it produces.

From (1.6), (1.7), (1.11a), and (1.22) we have the amplitude in the velocity as

$$v_m = \frac{F_m}{\sqrt{\left(\omega_e M - \frac{\chi}{\omega_e}\right)^2 + r^2}}. \tag{1.23}$$

Figure 8 shows this.

The mechanical impedance is then

$$\zeta = \frac{F_m}{v_m} = \sqrt{\left(\omega_e M - \frac{\chi}{\omega_e}\right)^2 + r^2}. \tag{1.24}$$

Figure 9 shows this as a function of frequency for several r; the impedance increases with the rigidity, and a given force produces an amplitude in the velocity inversely related to the impedance. This impedance has the same dimensions as resistance (dyne-sec/cm or kg-sec/m).

The impedance is minimal at resonance, where it equals the resistance r.

Mechanical impedance can be applied also to more complex mechanical systems, including ones with distributed constants (see below). The points in a composite system may have various velocities, so the impedance is defined by reference to the velocity acquired by the point to which the force is applied.

All the above arguments apply only to linear systems (ones whose mass, losses, and elasticity are not dependent on the force), which are described by linear differential equations with constant coefficients. Other systems have nonlinear equations, and the oscillations in these are much more complicated.

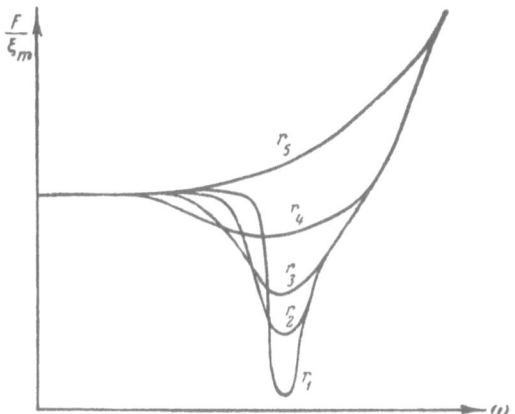

Fig. 7. Acoustic rigidity as a function of frequency for several values of the mechanical loss ($r_1 < r_2 < r_3 < r_4 < r_5$).

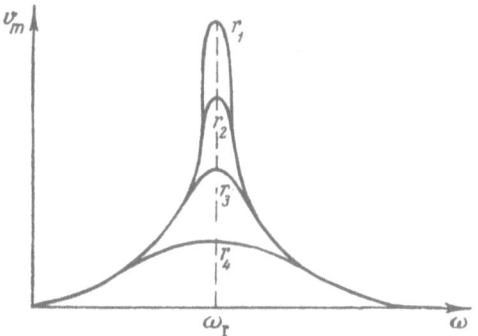

Fig. 8. Resonance curves for the amplitude in the velocity ($r_1 < r_2 < r_3 < r_4$).

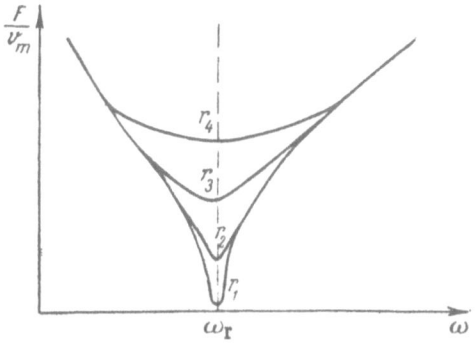

Fig. 9. Mechanical impedance as a function of frequency ($r_1 < r_2 < r_3 < r_4$).

Strictly speaking, an ultrasonic machining system is nonlinear, for its motion is affected by impact with the surface; the elasticity alters greatly at the moment of impact. However, it is found that the degree of nonlinearity is not high, so no great error is involved in the use of a linear approximation.

2. Propagation of Elastic Waves in Liquids and Gases

The oscillations excited at a point in any material system propagate as elastic waves (waves of alternating compression and tension). The speed of propagation is governed by the parameters of the medium (Table 1); it is not dependent on the amplitude or frequency, except in certain cases.

The most important characteristic of the wave is the distance traveled in one cycle (distance between adjacent compressions), which is called the wavelength λ; this is related to the velocity c and frequency f:

$$\lambda - \frac{c}{f}. \tag{1.25}$$

The particles in a gas or liquid confined in an infinite tube vibrate along the direction of propagation; the wave is plane, for its fronts form planes perpendicular to the direction of propagation.

The velocity potential φ may be used to describe the process; this quantity has no physical substratum, but it enables us to characterize the process at any point at any instant. This is related to the velocity of the particles as follows:

$$v = -\frac{\partial \varphi}{\partial x}, \tag{1.26}$$

in which x is a coordinate along the direction of propagation.

The points of compression have a pressure somewhat above the static pressure; this acoustic pressure is related to the velocity potential by

$$p = \rho_0 \frac{\partial \varphi}{\partial t}, \tag{1.27}$$

in which ρ_0 is the density of the medium at rest.

The velocity potential may be deduced from the wave equation, which for a plane wave takes the form

$$\frac{\partial^2 \varphi}{\partial x^2} = \frac{1}{c^2} \frac{\partial^2 \varphi}{\partial t^2}, \tag{1.28}$$

in which x is the coordinate and c the velocity as above. The wave may propagate from right to left or vice versa, so the general integral of (1.28) has two terms:

9

TABLE 1. Speeds of Propagation and Wave Impedances for Gases and Liquids

Medium	Density ρ, g/cm^3	Velocity c, cm/sec	Wave impedance, g/cm^2-sec	Temp., °C
Air	$1.29 \cdot 10^{-3}$	$0.343 \cdot 10^5$	44.2	20
Oxygen	$1.33 \cdot 10^{-3}$	$0.328 \cdot 10^5$	43.6	20
Nitrogen	$1.17 \cdot 10^{-3}$	$0.351 \cdot 10^5$	41.0	20
Carbon dioxide	$1.85 \cdot 10^{-3}$	$0.248 \cdot 10^5$	45.8	20
Acetone	1.10	$1.06 \cdot 10^5$	$1.17 \cdot 10^5$	20
Benzene	0.878	$1.32 \cdot 10^5$	$1.17 \cdot 10^5$	20
Water	0.997	$1.497 \cdot 10^5$	$1.49 \cdot 10^5$	25
Mercury	13.595	$1.451 \cdot 10^5$	$1.972 \cdot 10^5$	20
Toluene	0.866	$1.328 \cdot 10^5$	$1.15 \cdot 10^5$	20
Ethanol	0.789	$1.180 \cdot 10^5$	$0.93 \cdot 10^5$	20
Spindle oil	0.905	$1.342 \cdot 10^5$	$1.21 \cdot 10^5$	32
Kerosene	0.825	$1.295 \cdot 10^5$	$1.07 \cdot 10^5$	34
Transformer oil	0.895	$1.425 \cdot 10^5$	$1.28 \cdot 10^5$	32.5

$$\varphi = \varphi_{m_1} e^{j\omega\left(t-\frac{x}{c}\right)} + \varphi_{m_2} e^{j\omega\left(t+\frac{x}{c}\right)}, \tag{1.29}$$

in which φ_{m_1} is the amplitude in the potential for the wave from left to right, φ_{m_2} is the same for the other wave, and e is the base of Naperian logarithms.

The sign in front of the term containing x indicates the direction of propagation.

The boundary conditions define φ_{m_1} and φ_{m_2}; in the simplest case there is only one wave (e.g., from left to right), when

$$\varphi = \varphi_{m_1} e^{j\omega\left(t-\frac{x}{c}\right)} \tag{1.30}$$

or, in simpler form,

$$\varphi = \varphi_{m_1} \begin{matrix} \sin \\ \cos \end{matrix} \omega\left(t-\frac{x}{c}\right). \tag{1.30a}$$

The pressure and velocity are given by (1.26) and (1.27) as

$$v = \varphi_{m_1} \frac{\omega}{c} \cos \omega\left(t-\frac{x}{c}\right), \tag{1.31}$$

$$p = \varphi_{m_1} \rho_0 \omega \cos \omega\left(t-\frac{x}{c}\right). \tag{1.32}$$

Here $\varphi_{m_1} \omega/c = v_m$ is the amplitude in the velocity and $\varphi_{m_1} \rho_0 \omega = p_m$ is the amplitude in the acoustic pressure.

These expressions show that the velocity and pressure have a double periodicity in a traveling wave (in time and space). For t = constant we have the velocity and pressure as functions of position along the direction of propagation; this moment may be chosen arbitrarily, so we can put t = 0, which gives these distributions as

$$v(x) = v_m \cos \frac{\omega}{c} x, \tag{1.31a}$$

$$p(x) = p_m \cos \frac{\omega}{c} x. \tag{1.32a}$$

Conversely, we can put x = 0 in (1.31) and (1.32), which gives us the velocity and pressure as functions of time at a fixed point:

$$v(t) = v_m \cos \omega t, \tag{1.31b}$$

$$p(t) = p_m \cos \omega t. \tag{1.32b}$$

The velocity and displacement are related by the $v = \partial \xi / \partial t$ of (1.7), so the amplitude of the displacement is readily found in terms of the velocity potential:

$$\xi_m = \frac{v_m}{\omega} = \frac{\phi_m}{c}. \tag{1.33}$$

At any cross section in this tube there is an alternating force, whose amplitude is given by

$$F_m = p_m S, \tag{1.34}$$

in which S is the area of the wave front (here the cross section of the tube). The analogy with the vibrating system enables us to introduce a mechanical impedance z as the ratio of the force to the velocity:

$$z = \frac{F_m}{v_m} = \frac{S p_m}{v_m}. \tag{1.35}$$

Substituting for p_m and v_m from (1.31) and (1.32), we have

$$z = S \rho_0 c, \tag{1.36}$$

which shows that the impedance is independent of the amplitude and frequency, being governed solely by S and the properties of the medium.

The dimensions of this impedance are dyne-sec/cm (cgs) or kg-sec/m (MKS).*

The $\rho_0 c$ appearing in (1.36) is essentially the specific impedance (impedance per unit area); it is called the wave impedance and is an important parameter. We see from (1.35) and (1.36) that

$$\rho_0 c = \frac{p_m}{v_m}. \tag{1.37}$$

The pressure and velocity always vary in phase, as (1.31) and (1.32) show; only traveling plane waves have this feature, and the general case is one of a phase difference ψ between the two. This difference is important in the derivation of the power carried by the wave (energy flux), which can be shown to be

$$W_a = \frac{1}{2} S p_m v_m \cos \psi. \tag{1.38}$$

For a plane wave, for which $\psi = 0$,

$$W_a = \frac{1}{2} S p_m v_m. \tag{1.39}$$

The quantity W_a/S (density of energy flux) is called the intensity, which for a plane wave is

$$I = \frac{W_a}{S} = \frac{1}{2} p_m v_m. \tag{1.40}$$

The unit of intensity is erg/cm²-sec (cgs) or W/cm² (practical system).

The total power and intensity may be put in rather different form by the use of (1.37):

$$W_a = \frac{1}{2} S \frac{p_m^2}{\rho_0 c} = \frac{1}{2} S v_m^2 \rho_0 c, \tag{1.39a}$$

---

* The mechanical impedance is sometimes replaced by the acoustic impedance, which is defined as the ratio of the force to the volume (not linear) velocity. The mechanical impedance is of greater value in the cases to be considered here.

11

$$I = \frac{1}{2}\frac{p_m^2}{\rho_0 c} = \frac{1}{2}v_m^2 \rho_0 c. \qquad (1.40a)$$

If the wave strikes an interface with another medium, only part of the energy enters that medium, the rest being reflected. The ratio of the reflected power to the incident power is the reflection coefficient β, which for a plane wave striking the interface at right angles is defined by the mechanical impedances z_1 and z_2 of the two media:

$$\beta = \left(\frac{z_1 - z_2}{z_1 + z_2}\right)^2. \qquad (1.41)$$

The case $\beta = 0$ corresponds to equal impedances; the greater the difference between the mechanical impedances (or wave ones, since we consider a wavefront whose area is not affected by reflection) the greater the coefficient. For instance, $\beta \approx 1$ for an air-water interface, for $\rho_0 c$ for air is 42 as against 1.5×10^5 for water. The reverse occurs with water and carbon tetrachloride, which differ in density (1.0 and 1.5) and velocity (1500 and 1000 m/sec) but which have equal impedances, so $\beta = 0$. The energy passes through without loss, and the transmission coefficient is $1 - \beta$.

A wave striking the interface at an angle gives rise to refracted and reflected waves that obey the laws of geometrical optics: the angle of incidence is equal to the angle of reflection, and the ratio of the sine of the angle of incidence to the sine of the angle of refraction is equal to the ratio of the velocities of propagation.

The ratio of the transmitted energy to the reflected energy is governed by the angle of incidence as well as by the impedances, which also govern the angle of refraction via the velocities.

3. Propagation of Elastic Waves in Solids

The above longitudinal waves can also exist in solids; in this case the alternating pressure is replaced by an alternating stress (Fig. 10a). Solids can also carry other types of wave; an infinite medium can transmit transverse (shear) waves (Fig. 10b), whose velocity of propagation is always below that for longitudinal waves (Table 2); this is controlled by the shear modulus, whereas that of a longitudinal wave is governed by the bulk modulus (Table 3). Both types of wave can occur independently.

A finite body in addition can support waves involving bending (Fig. 10c); there are also others such as compressional waves in rods.

Waves in a rod whose transverse dimensions are small (Fig. 10d) are influenced by the Poisson effect (transverse expansion in response to longitudinal compression), which is characterized by the ratio of these, which is denoted by μ and varies from 0.2 to 0.35.

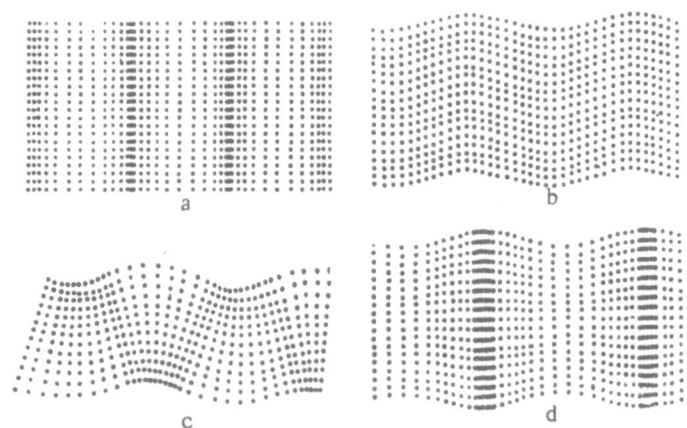

Fig. 10. Waves in a solid: a) compression waves; b) shear waves;
c) bending waves; d) compression waves in a thin rod.

TABLE 2. Elastic Constants and Propagation Velocities for Some Solids at 20°C

Material	Density ρ, g/cm^3	Young's modulus E, kg/mm^2	Shear modulus G, kg/mm^2	Poisson's ratio μ	Wave speed c in rod, m/sec	Wave speed c_∞ in unbounded volume, m/sec	Shear-wave speed, m/sec	Wave imp, $\rho_0 c$, g/cm^2-sec
Aluminum	2.7	7,100	2,640	0.34	5080	6260	3080	$13.7 \cdot 10^5$
Tungsten	19.1	36,200	13,400	0.35	4310	5460	2620	$82.0 \cdot 10^5$
Iron	7.8	21,000	8,200	0.28	5170	5850	3230	$40.4 \cdot 10^5$
Brass	8.8	16,600	6,250	0.33	4300	5240	2640	$37.4 \cdot 10^5$
Nickel	8.8	20,540	7,850	0.31	4785	5630	2960	$41.5 \cdot 10^5$
Lucite	1.18	535	151	0.35	2140	2670	1121	$2.5 \cdot 10^5$
Porcelain	2.41	5,860	2,381	0.23	4884	5340	3120	$11.7 \cdot 10^5$

TABLE 3. Types and Speeds of Waves in Solids

Wave	Medium	
	unbounded ($\lambda \ll$ dimensions)	bounded ($\lambda \gg$ dimensions)
Pure longitudinal (Fig. 12a)	$c_\infty = \sqrt{\dfrac{E}{\rho}\dfrac{1-\mu}{(1+\mu)(1-2\mu)}}$	Do not exist
Pure transverse (shear; Fig. 12b)	$c_\tau = \sqrt{\dfrac{G}{\rho}}$	Do not exist
Bending waves (Fig. 12c)	Do not exist	Rod of radius r $$\sqrt{\pi r f}\ \sqrt[4]{\dfrac{E}{\rho}}$$ Plate of thickness d $$\sqrt{\pi d f}\ \sqrt[4]{\dfrac{E}{3(1-\mu^2)\rho}}$$
Rod (Fig. 12d)	Do not exist	$c = \sqrt{\dfrac{E}{\rho}}$
Torsion	Do not exist	Thin rod $$c_r = \sqrt{\dfrac{G}{\rho}}$$
Surface waves	$c_s = \dfrac{0.87+1.12\mu}{1+\mu}\sqrt{\dfrac{G}{\mu}}$	

E = Young's modulus, ρ = density, G = shear modulus, μ = Poisson's ratio, f = frequency.

The transverse effects arising from this effect in a longitudinal wave are not apparent for an infinite body (infinite wave front), because the effects in adjacent parts are in opposition (the individual areas experience hydrostatic compression). A thin rod allows this lateral expansion, so traveling swellings occur (Fig. 10d). This makes the material effectively of lower rigidity; the deformation is governed by Young's modulus, which is always lower than the bulk modulus. Propagations in an unbounded solid and in a thin rod represent extreme cases; a rod of diameter neither large nor small relative to the wavelength shows complicated effects, and the velocity of propagation lies between c_∞ and c.

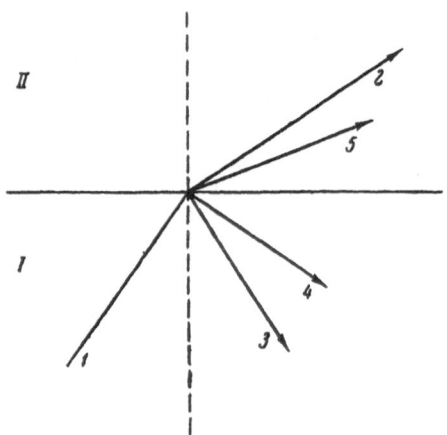

Fig. 11. Reflection at the interface between solids I and II: 1) incident longitudinal wave; 2) refracted longitudinal wave; 3) reflected longitudinal wave; 4) reflected shear wave; 5) reflected shear wave.

Bending and torsional waves can occur in rods, and longitudinal radial waves also (if the radius is larger than the wavelength). Surface waves can propagate on any free surface.

Only the bending waves have velocities dependent on the wavelength (frequency). Table 3 gives data and formulas, while Table 2 gives the elastic constants for the main wave types for the most commonly used materials. *

These waves can occur singly or in combination, so the wave patterns can be complicated.

Any change in the propagation conditions, such as reflection at an interface, may cause one wave type to change to another. For instance, a plane longitudinal wave may fall on an interface; then the reflected and transmitted longitudinal waves (which occur in liquids) are accompanied by two shear waves (one in each medium), whose directions are governed by the velocities in those media (Fig. 11).

Most of the applications in machining employ longitudinal waves in rods whose diameters are less than the wavelength. All other types of wave are to be reckoned as parasitic, for they may interfere with the machining as well as weaken the wanted waves. For instance, bending vibrations may be excited by slight curvature or inhomogeneity in the rod; these transverse vibrations cause a loss of accuracy in the machining. Particularly hazardous cases are those in which the natural frequencies for unwanted types lie near the working frequency (see below). The subsequent discussion relates to longitudinal waves in rods.

Some of the energy in a wave is lost on account of internal friction (e.g., from viscosity); the process is one of absorption, and the absorbed energy is converted to heat. The intensity then falls off with distance. The absorption of a plane wave is described by

$$I_x = I_0 e^{-2\alpha x}, \tag{1.42}$$

in which I_0 is the intensity at the initial point, I_x is the intensity at a distance x from that point, and α is the absorption coefficient.†

The last is a constant of the material, which is usually proportional to the frequency for the range 2-100 kc. Table 4 gives some values.

TABLE 4. Attenuation of Longitudinal Waves in Rods of Materials

Material	ρ, g/cm^3	c, m/sec	$\dfrac{\alpha}{f} \cdot 10^{-9}$, sec/cm	Q
Aluminum	2.68	5130	0.61	10,000
Magnesium	1.7	5100	1.08	5,700
Tungsten steel	8.52	4720	0.38	8.100
Molybdenum steel	8.39	4700	1.42	4,700
Fused quartz	2.2	5110	1.23	5,000
Pyrex	2.32	5350	4.89	1,200

* See note on p. 15.

† The absorption coefficient is sometimes given as $\alpha' = 2\alpha$; to avoid misunderstanding, this is pointed out here, for the present notation is more convenient. Here α gives the reduction in the stress or deformation over a distance of 1 cm, while α' gives the reduction in the intensity, which is not measured directly.

TABLE 5. Physical Constants of Metals and Alloys Used in Vibrators

Use	Material	Density ρ, g/cm³	Young's modulus E, dyne/cm²	Velocity of sound c, cm/sec	ρc, g/cm²·sec	Elastic limit, dyne/cm²	Yield point, dyne/cm²	Poisson's ratio μ	Saturation magnetostriction λ_s
Magnetostriction	Nickel	8.9	$2.06 \cdot 10^{12}$	$4.76 \cdot 10^{5}$	$42.4 \cdot 10^{5}$	$2 \cdot 10^{8}$	$46 \cdot 10^{8}$	0.31	$35 \cdot 10^{-6}$
	Alfer Yu-14	6.65	$1.73 \cdot 10^{12}$	$5.10 \cdot 10^{5}$	$33.9 \cdot 10^{5}$	$55 \cdot 10^{8}$	$77 \cdot 10^{8}$	–	$40 \cdot 10^{-6}$
	Permendur K49F2 ...	8.08	$2.14 \cdot 10^{12}$	$5.15 \cdot 10^{5}$	$41.6 \cdot 10^{5}$	–	$44 \cdot 10^{8}$	–	$70 \cdot 10^{-6}$
	K-65 alloy	8.25	$2.24 \cdot 10^{12}$	$5.20 \cdot 10^{5}$	$42.9 \cdot 10^{5}$	–	$66 \cdot 10^{8}$	–	$90 \cdot 10^{-6}$
Structural parts	Aluminum	2.70	$0.70 \cdot 10^{12}$	$5.24 \cdot 10^{5}$	$14.1 \cdot 10^{5}$	$1.5 \cdot 10^{8}$	$9 \cdot 10^{8}$	0.33	–
	Duralumin	2.70	$0.73 \cdot 10^{12}$	$5.20 \cdot 10^{5}$	$14.0 \cdot 10^{5}$	$15 \cdot 10^{8}$	$35 \cdot 10^{8}$	–	–
	Iron	7.86	$2.00 \cdot 10^{12}$	$5.17 \cdot 10^{5}$	$40.4 \cdot 10^{5}$	$12 \cdot 10^{8}$	$25 \cdot 10^{8}$	0.28	$10 \cdot 10^{-6}$
	Brass	8.50	$1.00 \cdot 10^{12}$	$3.42 \cdot 10^{5}$	$29.1 \cdot 10^{5}$	$5 \cdot 10^{8}$	$40 \cdot 10^{8}$	0.37	–
	Copper	8.93	$1.20 \cdot 10^{12}$	$3.58 \cdot 10^{5}$	$32.0 \cdot 10^{5}$	$1.5 \cdot 10^{8}$	$40 \cdot 10^{8}$	0.35	–
	Steel	7.80	$2.10 \cdot 10^{12}$	$5.05 \cdot 10^{5}$	$39.3 \cdot 10^{5}$	$20 \cdot 10^{8}$	$40 \cdot 10^{8}$	0.18	–

Shear waves usually have less attenuation; α is also independent of amplitude for small amplitudes, but nonlinear absorption occurs at high amplitudes, when α increases (at first slowly, but then increasingly rapidly). Not much is known about absorption in metals at high amplitudes; it is impossible to give any reliable data at present. The total energy flux is W = IS and so is proportional to the cross section of the rod.

The above applies only if the conditions are such that Hooke's law is obeyed; the stress must not exceed the limit of proportionality.

Table 5 gives the elastic constants for some materials commonly used in the ultrasonic sections of machines (materials for magnetostriction vibrators and also for concentrators and other parts*). The last column gives the maximal magnetostriction strain, which is considered in Chapter 3.

4. Waveguides

Waves are commonly transmitted along systems whose lengths are much greater than their transverse dimensions; such systems are called waveguides. These prevent lateral leakage and guide the wave in a definite direction. A rod is a particular case of a waveguide. A waveguide is uniform if its properties and cross section do not vary along its length. Consider a plane wave propagating in a guide of cross section S and infinite length; then the mechanical impedance at any point is given by (1.36) as

$$z = S \rho_0 c. \qquad (1.43)$$

The input impedance is defined as the ratio of the force acting on the input to the velocity produced there; now (1.43) applies to any point in a uniform guide, so it applies to the input:

$$z_i = z = S \rho_0 c. \qquad (1.44)$$

The input impedance is then independent of frequency and amplitude.

The power entering the guide from a force of amplitude F_i uniformly distributed over the end is

$$W = \frac{1}{2} \frac{F_i^2}{z_i}. \qquad (1.45)$$

This shows that the power taken up from a given force is inversely proportional to the input impedance.

---

* The results in Tables 2, 4, and 5 were derived by various workers under various conditions, which explains the discrepancies. No attempt has been made to eliminate these.

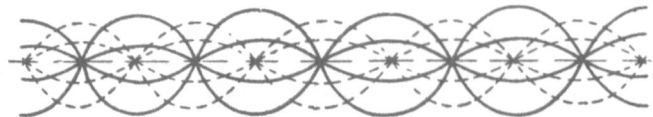

Fig. 12. Standing waves in a guide with free ends.

This means that a guide designed to transmit power to the tool must have a low input impedance, while a part designed to act as an isolating support must have a high impedance.

Practical guides are always of finite length, in which case the result is dependent on the acoustic load at the far end; if this is a mechanical resistance numerically equal to z, all the power passes to the load. The guide then carries a traveling wave, and its impedance remains $S\rho_0 c$. The guide is then matched to the load.

The result is different if the load is not matched. Consider first the case of no load; then all the energy reaching the far end is reflected and so is represented by the second term in (1.29), which we previously neglected. The complete picture is described by the expression

$$\varphi = \varphi_{m_1} e^{-j\omega\left(t-\frac{x}{c}\right)} + \varphi_{m_2} e^{-j\omega\left(t+\frac{x}{c}\right)}. \tag{1.29}$$

All the energy is reflected, so the amplitude of the potential for the reflected wave equals that for the incident one:

$$\varphi_{m_2} = \varphi_{m_1} = \varphi. \tag{1.46}$$

If now the length of the guide is an integral number of half-wavelengths,

$$n\frac{\lambda}{2} = l, \tag{1.47}$$

the guide will contain standing waves (Fig. 12). The full lines indicate the distribution of displacement and velocity; the broken ones, that of stress. The points at which the stress (displacement) is always zero are called nodes. For example, free ends are always stress nodes. The distance between adjacent nodes is half the wavelength. The points at which the displacement or stress is maximal are called antinodes; a free end is a displacement antinode. The distance between adjacent antinodes is also half a wavelength.

The input impedance of such a guide is zero, for the impedance is the ratio of the force to the velocity, while here the force is zero for a finite velocity. This means that an infinitely small force will cause the velocity to rise without limit [see (1.35)], as will the power drawn [see (1.45)].

This is impossible, of course, for a real system has losses, which restrict the growth. In this respect standing waves are very similar to the resonance in the system with one degree of freedom (section 1). The analogy can be extended, for a slight change in frequency or wavelength causes a large reduction in amplitude, on account of deviation from condition (1.47)

$$l = n\frac{\lambda}{2}.$$

We may say that the guide exhibits resonance; the natural frequency is given by (1.47) and (1.25), while the magnification is governed by the losses (Table 4):

$$Q = \frac{\pi}{c}\frac{f}{\alpha}.$$

The last column in Table 4 gives Q, which is a constant of the material for the above conditions (small amplitude, range 2-100 kc), being independent of frequency. The peak becomes higher and sharper as the magnification increases.

The impedance of a resonant guide at a velocity or displacement node is infinitely great, for any force at this point cannot produce a finite velocity. This means that the guide may be clamped at such a point without loss of energy.

16

In the general case the guide works into a load, which may be purely resistive. The pattern is then intermediate between those for pure traveling waves and pure standing ones. The resultant traveling wave transmits part of the power from the source to the load; the amplitude of this wave and the proportion of the power transmitted are governed by the degree of matching, i.e., by $R/S\rho_0 c$. The balance of the energy is present in the guide as standing waves; the guide remains resonant but the magnification falls, for the natural loss is accompanied by transfer of power to the load.

The traveling-wave coefficient (TWC) specifies the relation of traveling to standing waves; the TWC is zero if there is no traveling wave, and no power is transmitted along the guide. It is unity if there is no standing wave, which implies perfect matching (all power passes to the load). This is considered in more detail in Chapter 3.

Any deviation from uniformity (change in material or sudden change in cross section) can cause reflection; (1.43) shows that both factors affect the mechanical impedance, which means that some fraction of the energy is reflected. The reflection coefficient is given by (1.41) as

$$\beta = \left(\frac{z_1 - z_2}{z_1 + z_2}\right)^2.$$

If the cross section alone alters, we have from (1.41), since $z = S\rho_0 c$,

$$\beta = \left(\frac{z_1 - z_2}{z_1 + z_2}\right)^2 = \frac{\frac{S_1}{S_2} - 1}{\frac{S_1}{S_2} + 1} = \left(\frac{m-1}{m+1}\right)^2, \tag{1.41a}$$

in which $m = S_1/S_2$ (S_1 is the cross section before the point of narrowing; S_2 is that after it).

For instance, the diameter may be reduced by a factor 3; then the area is reduced by a factor 9, and (1.41a) with $m = 9$ gives $\beta = 0.64$, so only about one third of the energy is transmitted. Similarly, a change by a factor 4 gives $m = 16$, for which $\beta = 76\%$.

This question is considered in more detail below, for a guide of changing cross section is, strictly speaking, not uniform, and a discussion merely in terms of a junction between two guides fails to bring out all the features.

The load may have a reactive component, on account of a mass or elastic member linked to the end. A reactive load affects the resonance frequency and so alters the standing-wave condition.

Only a small proportion of the energy is transmitted to the load in ultrasonic cutting, so the systems show pronounced resonances.

Important elements in such systems are nonuniform guides (rods whose cross sections vary along their lengths). Chapter 3 deals with the calculation of components of such systems.

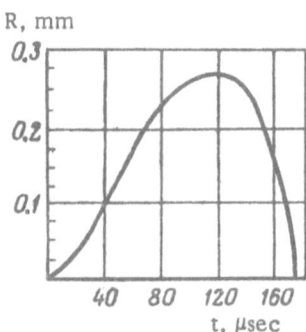

Fig. 13. Oscillation
of a cavitation bubble.

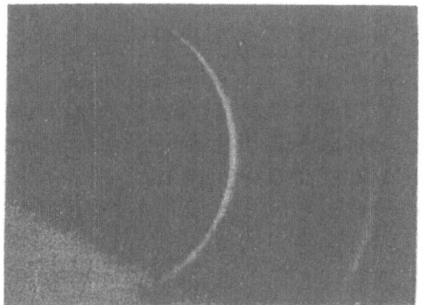

Fig. 14. Photograph of the shock wave
produced by collapse of a cavitation
bubble.

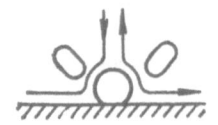

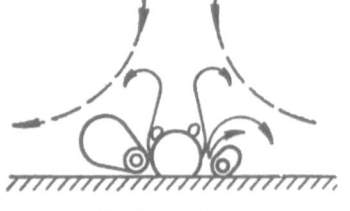

Fig. 15. Modes of formation of microflows near an oscillating bubble.

5. High-Intensity Waves in Liquids

Weak waves (ones of the order of 10^{-9} W/cm^2, normal speech levels) cause no residual effects in liquids; all compressions and so on vanish as soon as the wave has passed. Intensities of the order of 1 W/cm^2 produce several secondary effects, some of which persist after the wave has passed.

One of the most important secondary effects is cavitation (formation of small gaps as a result of the tensile stresses produced during the phases of rarefaction). These gaps arise at the weakest points and very often take the form of small air bubbles, which derive from air on the surface of the solid (tool, workpiece, abrasive particle) at points of incomplete wetting.

Surface tension causes these bubbles to take a spherical shape; if the tensile forces persist after such a bubble has arisen, the bubble may expand to several times its initial diameter. The bubble shrinks as the forces decrease, and the rate of shrinkage may become very high, the result being hydraulic shock when the collapse is completed. Figure 13 shows the radius R as a function of time, from which we see that the rate of collapse is much higher than the rate of expansion. Figure 14 shows a shock wave · produced by collapse of a cavitation bubble. Calculations and measurements show that the pressures in such shock waves may be several hundred atmospheres, which may be sufficient to cause damage to adjacent solids. In addition, they can accelerate solid particles suspended in the liquid.

Cavitation bubbles are most commonly seen on the moving end of the tool in ultrasonic cutting; they are less common on particles of abrasive and least of all on the workpiece.

Cavitation plays a major part in ultrasonic cutting; it is considered in more detail in Chapter 2.

Another important effect is the production of flows of liquid where the ultrasonic fields are strong; some of these may spread throughout the liquid, and in this case they originate from the radiator. The detailed paths are much dependent on the shape and surface roughness of any solids encountered.

These large-scale flows may be accompanied by local microflows associated with the oscillations of cavitation (or other) bubbles attached to surfaces. The form of these microflows is dependent on the frequency, bubble size, density, and viscosity. Figure 15 shows some possible forms. The frequencies (10-40 kc) and amplitudes used in ultrasonic cutting are such that the velocities in these flows can be over 10 cm/sec for water, but this is the upper limit for more viscous liquids. Microflows can also occur near the edge of a vibrating solid.

These flows also play an important part in cutting, for they transport the suspended particles. They are considered in more detail in Chapter 2.

To conclude, we note a few of the relations between acoustic and ultrasonic vibrations with respect to the machining of hard and brittle materials.

The human ear can perceive the range 30 to 15,000 c/s; the upper limit (above which lies the ultrasonic range) is rather indefinite and varies greatly from one individual to another, and it also tends to fall with advancing age. Moreover, it is somewhat dependent on the state of the nervous system at the time of measurement. There is no essential physical difference between the upper acoustic range (5000 to 15,000 c/s) and the low ultrasonic range (15,000 to 30,000 c/s), for both obey the laws outlined above; in particular, both ranges can be used in machining.

The main reason for using ultrasonic frequencies is to provide satisfactory working conditions. Audible frequencies of the required intensities would be heard as extremely loud sounds, which would cause fatigue and might even produce irreversible changes in the auditory apparatus.

The problem is not completely solved by the use of ultrasonics alone, for fairly strong audiofrequency vibrations are produced during the cutting. In part these result from ringing in the tool resulting from the blows, and in part from the poorly understood process of subharmonic generation (production of frequencies divided by 2, 3, and so on) from cavitation. These audible frequencies are irritating.

Although ultrasonic frequencies are not audible, they are not unconditionally harmless to the ear. This aspect is under study. It may well be found that shielding to suppress audible and inaudible frequencies may be required for this reason.

Literature Cited

1. G. S. Gorelik, Vibrations and Waves, Moscow-Leningrad, Fizmatizdat, 1958.
2. I. I. Teumin, Ultrasonic Vibrating Systems, Moscow, Mashgiz, 1959.
3. L. Bergmann, Ultrasonics [Russian translation], Moscow, IL, 1956.

CHAPTER 2

PHYSICAL PRINCIPLES OF ULTRASONIC MACHINING

6. The Basic Experimental Laws in Ultrasonic Machining

In the early stages it was found [1-4] that the cutting rate* and surface finish are dependent on the frequency and amplitude of the vibration, on the static load, and on the hardness, grain size, and concentration of the abrasive. It was found that brittle materials are cut more rapidly.

Table 6 gives values found by Neppiras [8] for identical conditions (16.3 kc, amplitude 12.5μ, 100 mesh boron carbide), the rate of penetration for glass being 0.6 mm/min. Table 7 gives his results for the rate of cutting and the relative wear of the tool for some brittle materials; the values in parentheses are ones found by Hartley [5] under other conditions (38μ at 25 kc, 320 mesh boron carbide). The two sets of results [5, 8] sometimes differ.

Table 7 shows that plastic materials are cut much less rapidly than brittle ones; for instance, the rate for steel is somewhat less than that for tungsten carbide and 20-50 times less than that for glass, although steel is roughly as hard as glass but the carbide is much harder. On the whole, the rate decreases as the material becomes harder. There is no exact relation of rate to hardness and brittleness.

The process involves vigorous cavitation in the presence of abrasive and is very difficult to examine. Various forces are present in the gap between tool and workpiece [6], and these produce erosion. Among them we have mechanical impact of the tool on the particles of abrasive lying in the surface, impacts from moving particles, and shock waves from cavitation at particles or at the surface. The theories of the processes were at first neglected, because interest was directed to the cutting rate and surface finish as governed by the pressure between tool and workpiece, the composition of the abrasive suspension, the frequency, the vibration amplitude, and so on.

Nomoto [7] found that the force affects the cutting rate. He used a piezoelectric vibrator of sandwich type (Langevan) at 26.2 kc (amplitude 10μ), the tool being a drill or rod of diameter 1 to 6 mm. He measured the

TABLE 6. Relative Rates of Ultrasonic Machining of Metals

Material	Rate, mm^3/min	Rate as % of rate for glass
Brass	2.2	6.6
Tungsten	1.6	4.8
Titanium	1.35	4.0
Steel ($R_C^* = 62$)	1.3	3.9
KE672 Steel ($R_C = 66$)	4.7	1.4
Chrome steel	4.7	1.4

* R_C is the Rockwell hardness.

---

* By cutting rate C is meant the volume of material removed in unit time; this is proportional to the rate of penetration for a fixed cutting area S: $C = Sv$.

TABLE 7. Relative Rates of Ultrasonic Cutting of Materials

Material	Rate, mm³/min	Rel. rate, %	Wear as% of rate	Hardness, kg/mm²
Barium titanate	37	110	–	–
Soda glass	33.5 (320)	100 (100)	0.5	560
Borosilicate glass	29	86	–	600
Heavy flint	25	75	–	450
Porcelain	23.5	70	–	–
Crystal quartz	19 (140)	57 (44)	2	1100-1200
Ferrite	12.5 (240)	37 (75)	–	220-260
Crystal germanium	10.5 (240)	31 (75)	–	–
Synthetic sapphire	6.4	19	–	2700
Synthetic ruby	6.1 (65)	18 (19)	50	2800
Tungsten carbide	1.4 (29)	4.1 (9)	100	1800-2400
Tool steel	1.3 (22)	3.9 (7)	100	750

depth of the hole produced in glass with 320 mesh carborundum in 10 sec and found that there is a best force (Fig. 16), which was dependent on the size of the tool. Similar results were obtained for quartz, boron carbide, and hard alloys. The hole produced in glass was deeper at the edges (Fig. 17), and the glass bore radial scratches. He considered that the result was determined by the behavior (velocity) of the particles under the tool, the velocity being higher at the edges.

Neppiras [8, 9] made a detailed study; he related the rate and surface finish to the principal parameters and attempted to elucidate the nature of the damage. He considered that the most probable mechanism was essentially one of microchipping resulting from blows by particles of abrasive; he showed that cavitation erosion was unimportant. For instance, a gap of 0.12 mm between tool and glass gave rise to vigorous cavitation, but hardly any impression was made in 30 min; whereas a force of 1.3 kg at the same amplitude (25.1 μ) produced a rate of 2.5 mm/min. He performed many tests to relate the mean rate to the force, amplitude, and frequency, all tests being made on glass under fixed conditions with a tool 6.3 mm in diameter. Figure 18 shows the results.

The rate increases with the force at first, but there is later a fall, which he explained as follows. The stress pulse applied via the particles is proportional to the force when this is low, so the rate is proportional to the force for small forces. The motion of the abrasive becomes impaired at higher forces, so the rate falls.

The maximum rate and the optimum force increase with the amplitude. Similar results were obtained at three other frequencies (11, 16.3, and 19.5 kc).

The rate bears a square-law relation to the amplitude, as Fig. 19a shows for all four frequencies. The curves also show that the rate increases with the frequency for a fixed amplitude, which is to be expected, for

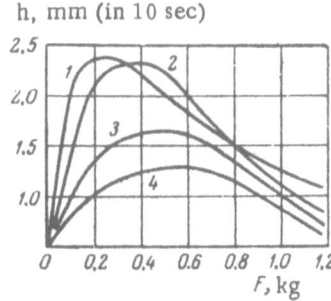

Fig. 16. Mean cutting rate as a function of force for tools of various diameters d [7].
1) 1 mm; 2) 2 mm; 3) 4 mm; 4) 6 mm.

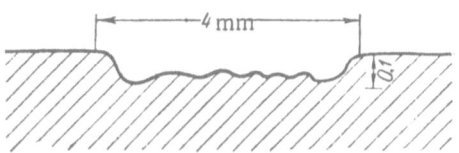

Fig. 17. Surface of glass after cutting [7].

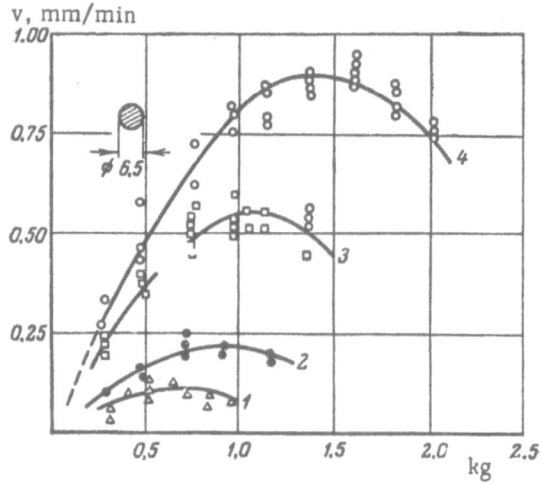

v, mm/min

Fig. 18. Cutting rate as a function of force at 5.1 kc and amplitudes $\xi_m(\mu)$ as follows: 1) 8; 2) 11; 3) 16.5; 4) 20.5.

the number of blows and the momentum of each both increase with the frequency, so the rate should be proportional to the square of the frequency. This is found to be so, but only for low frequencies, and the effect of frequency steadily decreases (Fig. 19b). The rate is given by $v = Cf^n$ at ultrasonic frequencies, in which n is less than one; a square-root relation fits the results very closely.

A need to examine the performance of thin tools arose from the application to the cutting of quartz, semiconductors, precious stones, and diamonds. D'yachenko et al. [10] found the relation of rate to force, amplitude, and tool thickness (Fig. 20); the form of the relation is not affected by the use of thin cuts. The best pressure and maximum rate increase with amplitude, as Neppiras found.

The rate is also governed by the hardness, particle size, and concentration of the abrasive; Fig. 21 shows the rate as a function of concentration for boron carbide and silicon carbide [8], which is proportional for low concentrations but becomes independent of concentration between 30 and 60% by volume. This relation was also reported by D'yachenko et al. [10]. Further studies have shown that the maximum rate is reached at much lower concentrations if the area of the tool or the force is reduced. The microhardness of the abrasive also affects the results; boron carbide (5000 kg/mm^2) is about 1.5 times as hard as silicon carbide (3000-3500 kg/mm^2), so the former gives a somewhat higher rate.

Coarser particles also cut faster; the rate is linearly related to the grain size for small grain sizes [8], but this ceases to apply when the grain size becomes comparable with the amplitude and the rate becomes constant. A high rate in conjunction with a specified surface finish is obtained only with a definite grain size, so many studies of the grain-size effect have been made [9-12]. It has been found that there is even a fall in rate above a certain grain size. Figure 22 [14] shows the relation for glass 3 mm thick. The optimal grain size is dependent on the amplitude and the tool thickness (Fig. 23) [9].

A tool 1.5 mm thick gave an optimal grain size of $4\xi_m$, but thinner tools did not give this amplitude relation, for the optimal size becomes dependent on the thickness of the tool rather than the amplitude (Neppiras had previously observed that the thickness must be at least 5 times the grain size). The amplitude relation is not in doubt, but it has not yet received a theoretical explanation.

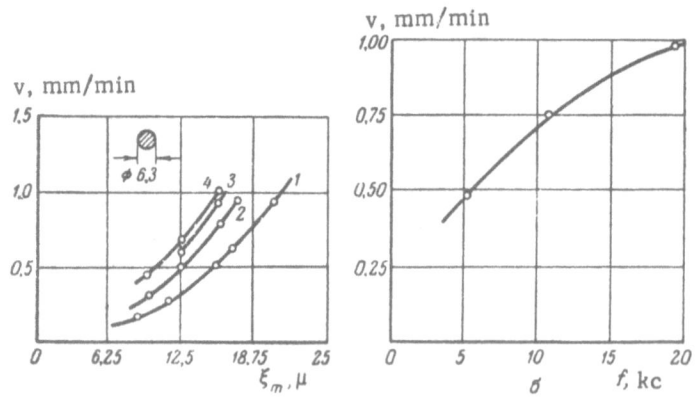

Fig. 19. Relation of maximum rate a) to amplitude and b) to frequency for a fixed amplitude of 15.7 μ [9]; frequencies (kc): 1) 5.1; 2) 11; 3) 16.3; 4) 19.5.

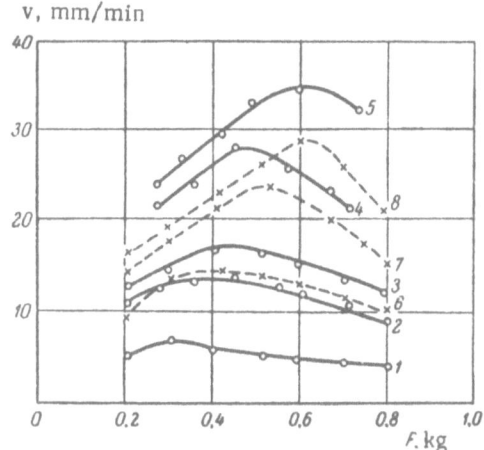

v, mm/min

Fig. 20. Rate as a function of force for various amplitudes and tool thicknesses [10]; the full lines are for a thickness of 0.2 mm, and the broken ones for 0.4 mm, for $\xi_m(\mu)$ as follows: 1) 21.5; 2) 29; 3) 33; 4) 39; 5) 46; 6) 27.5; 7) 36.5; 8) 42.

Cavitation plays no important part in the actual main cutting, but cavitation erosion can produce poor finish in the sides of the hole; cavitation currents are set up along certain parts of the tool, and particles of abrasive caught up by these give rise to grooves and hence poor surface finish. On the other hand, cavitation accelerates the cutting by providing vigorous circulation of abrasive under the tool. Liquids differ in their tendency to give cavitation. Other factors that affect the result are the density, viscosity, and surface tension. Neppiras [8] gave some results for common liquids (Table 8) as used with an amplitude of 16 μ with 320 grade boron carbide for small depths of cut. The values in parentheses are for cuts of depth 6.3 mm.

Figure 24 shows the relation of cutting rate to viscosity for glycerol-water mixtures, but the method of presentation is suspect, for properties other than the viscosity vary when glycerol is diluted with water.

The surface finish is important; it has been examined as a function of grain size and pressure [8], the former being much the more important. Figure 25 shows the mean height of the surface grooves and surface profiles as affected by grain size for glass and tungsten carbide. The finish improves as the grain size is reduced; tungsten carbide gives the better finish for a given grain size. The conclusion to be drawn is that the process is largely governed by the rate of abrasion by the particles (see also section 9).

The accuracy of machining is also important. Of course, it is inevitable that a hole is larger than the tool, on account of the need to feed the suspension into the cutting area; the gap grows larger as the hole deepens, the limit being set by the maximal size of the grains. The diameter of a blind hole at the bottom exceeds that of the tool by the diameter of the smallest grains. The gap is dependent [5] on the hardness of the material, on the grain size, on the force used, and on the duration of the operation. Moreover, wear on the surfaces inevitably makes the hole conical. Wear at the end reduces the length of the tool and so alters the resonance frequency. Ceramics have been used in tests with boron carbide No. 120 (pressure 0.15 kg/mm^2) and a tool in the form of a tube (outside diameter 7 mm, inside 4 mm) made of various materials [7]. Measurements were made of the wear (in length and diameter) for a given depth of cut (Table 9). The sequence as regards wear in length is not the

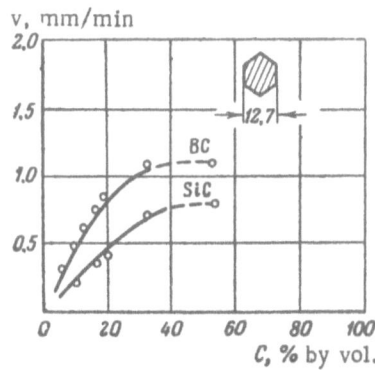

Fig. 21. Cutting rate as a function of concentration for abrasive of 100 mesh (boron carbide) or 220 mesh (silicon carbide) [8].

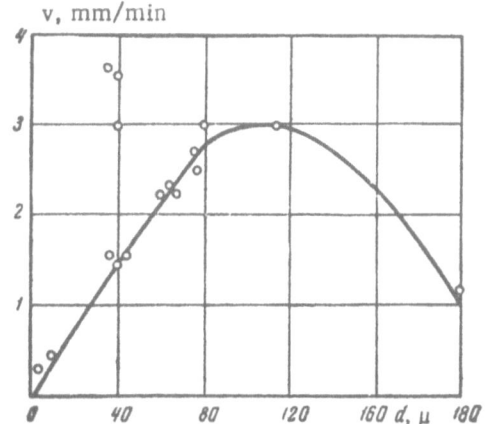

v, mm/min

Fig. 22. Mean cutting rate as a function of mean grain size [14,50].

TABLE 8. Rate of Cutting with Various Liquids

Liquid	Rate, mm/min	Viscosity, poise
Water	0.89 (0.51)	0.011
Benzine	0.63	0.006
Ethanol	0.51	0.012
Trichloroethylene	0.43	
Lubricating oil	0.31 (0.05)	0.5-10.0
Transformer oil	0.25	
Glycerol in water (% glycerol):		
100	0.01	0.36
75	0.13	0.15
67	0.31	0.07
50	0.45	0.02
0	0.99	0.01

same as that for wear in diameter, so there must be different mechanisms of wear. In fact, end wear is due to particles striking the end at right angles, whereas diameter wear is associated with tangential motion, so different properties must be involved.

D'yachenko et al. [10] have shown that the wear of the end face is dependent solely on the tool material and duration of use; various materials were used with a permendur plate 0.2 mm thick as tool (Table 10). The time taken to cut 1 mm into soda glass was taken as unity (4 sec), and the wear was 0.042 mm, so the wear per minute was 0.63 mm.

The results show that the interaction between tool and abrasive is not dependent on the properties of the workpiece; the cutting rate is governed by the abrasion produced by the grains.

Goetze [15] examine the effects of various factors on the cutting rate for SAE 1095 steel (0.9-1.05% carbon and 0.25-0.5% manganese). He used a tool 3.2 mm in diameter of the same steel.

Figure 26a shows the relation of cutting rate to amplitude for three frequencies for two grades of boron carbide; Fig. 26b shows the ratio of the rate to the amplitude as a function of frequency. The rate v is proportional to the product of ξ_m, f, and grain size d.

These results do not agree with Neppiras's [8] and with some later studies, because plastic deformation in steel is essentially different from brittle fracture.

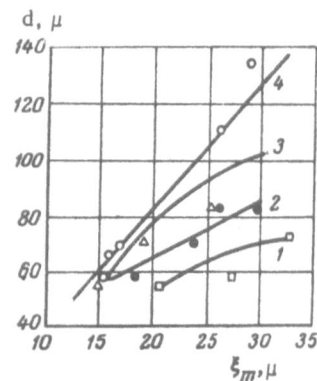

Fig. 23. Optimal value of mean grain size as a function of amplitude for tools of various thicknesses [10]: 1) 0.1 mm; 2) 0.2 mm; 3) 0.5 mm; 4) 1.5 mm.

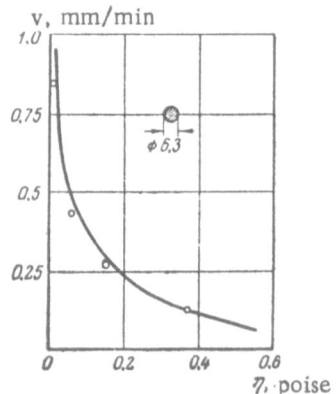

Fig. 24. Relation of cutting rate to viscosity [9].

TABLE 9. Effects of Tool Material on Wear with Ceramics Cut to a Depth of 15 mm

Material	End wear, mm	Diameter wear, mm
VK-8 hard alloy	0.20	0.031
R9 steel	0.27	0.068
U8 steel (ann.)	0.37	0.082
U8 steel (quenched)	0.41	0.045
45 steel	0.41	0.100
Kh18N9 steel	0.53	0.089

TABLE 10. Relative Wear of Tool in Machining of Various Materials

Material	Relative time to cut 1 mm	Relative wear	
		for 1 mm	per min
Soda glass	1.0	1.0	1.0
Optical glass	0.79	0.905	1.15
Quartz	1.82	2.14	1.11
Agate	3.12	2.88	0.92
Corundum	13	12	0.91
VK-8	20	17.3	0.87
Tungsten carbide	45	48	1.05

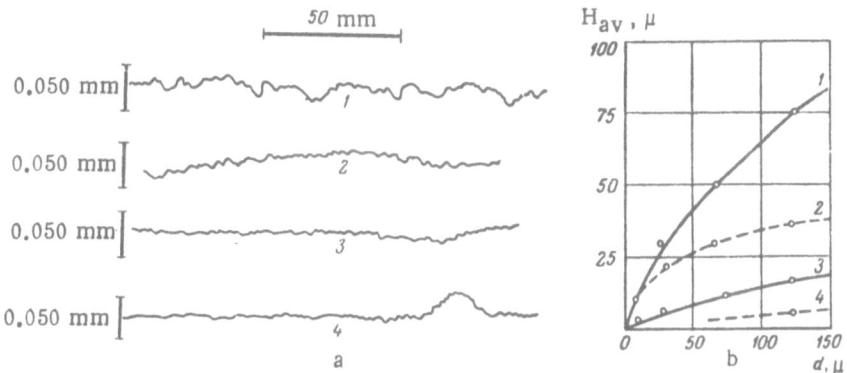

Fig. 25. Relation of surface finish to grain size [9]: a) profiles of glass worked with abrasives of the following grain sizes: 1) 80 mesh; 2) 320 mesh; 3) M14; 4) M7; b) relation of mean height H_{av} of grooves to mean grain size: 1 and 2 for glass, 3 and 4 for tungsten carbide. The full lines relate to the end and the broken lines to the side.

Goetze found that the rate is proportional to $\xi_m fd$, so he attempted to relate $v/\xi_m fd$ to the concentration of abrasive. From 374 measurements at 19, 25, and 43 kc with amplitudes from 10 to 50 μ and abrasives of mesh 100, 240, 400, and 600 he found that $v/\xi_m fd$ is proportional to concentration up to some limit. The limiting concentration was 8.4% for the 3.2 mm diameter used with a force of 3.1×10^6 dyne/cm². However, the results show a considerable spread; they have a gaussian distribution with a relative standard deviation of 0.236 (half the values deviate from the probable value by more than 30%). Goetze attributed these large discrepancies to random variations in the concentration of abrasive under the tool. It was later found [16] that the rate is related to the concentration for boron carbide of 320 mesh at 25 kc for the same pressure with a tool 1.6 mm in diameter. Here the maximum rate was reached at 25-30%, which was followed by a slight fall.

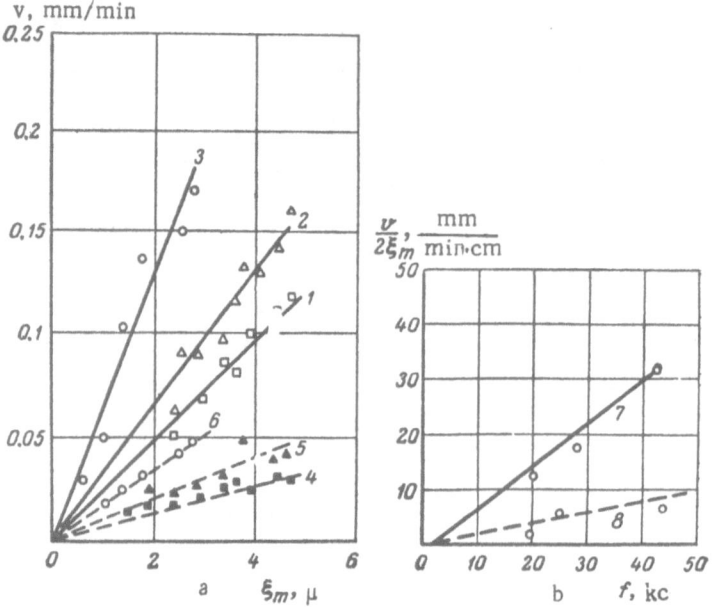

Fig. 26. a) Machining rate as a function of amplitude for several frequencies and grain sizes; b) ratio of rate to amplitude as a function of frequency [15]. The full lines are for boron carbide (240 mesh) and the broken ones for boron carbide M14: 1) 19.0 kc; 2) 24.7 kc; 3) 43.0 kc; 4) 19.0 kc; 5) 24.7 kc; 6) 43.0 kc; 7) boron carbide, 240 mesh; 8) boron carbide No. 14.

The above tests were done at constant pressure; the effects of the pressure were also examined in [17] with two tools of the same area but different perimeters (25 kc, 280 mesh, 50% by weight). Figure 27 shows that the rate is proportional to the pressure up to some limit, but thereafter there is no change. Circular tools gave the same result at various frequencies. Goetze found the following relations (Fig. 28) between the pressure, the length of perimeter, the area, the rate, the amplitude, the frequency, and the grain size:

$$ p = K \frac{l}{S} ; \qquad \frac{v}{2\xi_m f d} = M \frac{l}{S} . $$

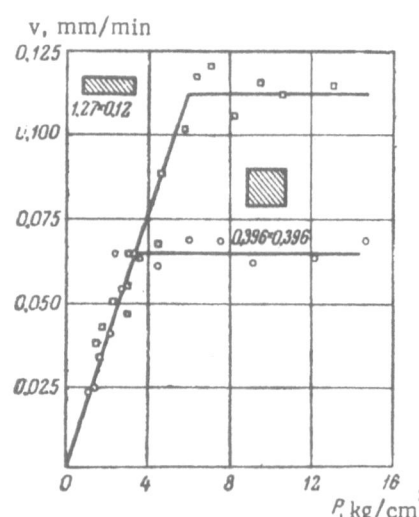

Fig. 27. Relation of rate to pressure for tools of different shapes [17].

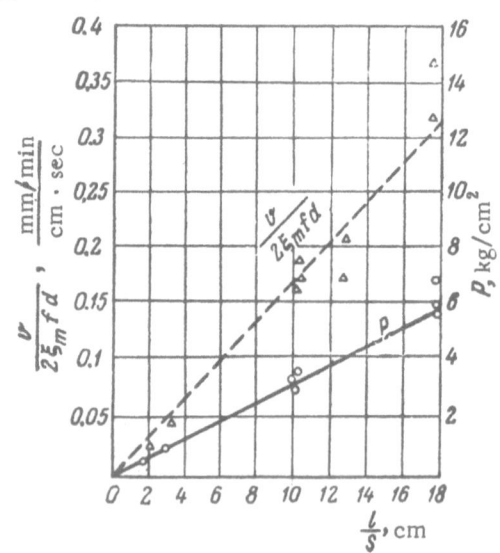

Fig. 28. Relation of relative rate and pressure to ratio of length of perimeter to area [17].

7. Theoretical Explanation of the Observed Laws

Miller [18] gave a theoretical semiquantitative discussion of Goetze's observations on the following basis: the particles are embedded in the workpiece and tool by the applied force, and this causes plastic deformation and work hardening. The hardened parts are removed by chipping. The rate was related to the basic parameters on the assumption that the rate is the product of the total volume of plastically deformed material by the work of hardening per unit volume and by the volume removed by chipping.

The plastically deformed volume was deduced as follows. The force Φ acting on a particle is $\Phi = F/N$, in which F is the force on all N particles. Each particle was assumed to be a cube of side d, and the depth of penetration was taken as proportional to the stress $\sigma = F/Nd^2$, to d, and to the reciprocal of the product of the shear modulus G by the Burgers vector b.* From this Miller found that the volume of plastically deformed material produced per second (i.e., per f blows) was

$$V_p \propto \frac{\sigma d}{Gb} d^2 fN = f \frac{dF}{Gb}.$$ (2.1)

On the assumptions that the work of hardening is proportional to the deformation rate (i.e., to f) and that the rate of removal is proportional to d^3, to N, and to f, Miller obtained for the volume removed in unit time

$$V' \propto \frac{fdF}{Gb} fd^3 fN.$$ (2.2)

The number of particles producing chipping (on the assumption that all particles under the tool do this) may be found as follows. A region of reduced pressure is produced under the tool as it recedes, and abrasive begins to flow into this; but the suspension is unable to fill all the space in the half-cycle, and the part that has entered is compressed as the tool moves back, some of it being pushed out. Cutting therefore starts at the edge. Eventually the groove becomes deep enough for the suspension to persist there throughout the cycle, whereupon cutting starts in the adjacent ring. The process gradually extends over the surface.

Miller assumed that the number of particles under the tool did not vary and was the number penetrating under during one vibration. This may be found from the equation of motion for a volume V_c having a mean density

$$\rho = \rho_a \frac{C+1}{C + \dfrac{\rho_a}{\rho_l}}$$

(C is the concentration; ρ_a and ρ_l are the densities of abrasive and liquid) in response to the atmospheric pressure p_0:

$$N = \frac{\pi r \xi_m p_0 C}{2f^2 \rho_a V_c d^3 (C+1)},$$ (2.3)

in which ξ_m is amplitude and r is radius of the tool. This is inserted in (2.2) to give the cutting rate as

$$v \propto \frac{\xi_m F fd}{Gb} \frac{C}{C+1}.$$ (2.4)

But the rate does not increase without limit as the force is raised; there must be an optimal value, because the particles become completely embedded in the workpiece at a stress σ_w or in the tool at a stress σ_t. The stress σ_0 needed to embed to a depth d is

$$\sigma_0 = \frac{2\sigma_w \sigma_t}{\sigma_w + \sigma_t}.$$ (2.5)

There is also a stress σ_a at which the grain cracks. The optimum pressure is then given by the smaller of σ_0 and σ_a multiplied by the number of particles under the tool:

---

* Pugh (Phil. Mag., Ser. 7, 45:824, 1954) derived this relation of deformation to stress from dislocation theory. The vector characterizes the lattice-distortion energy.

$$P_o = (\sigma_0 \text{ or } \sigma_a) \frac{\xi_m p_0 C}{2\rho_a \; rfd\,(C+1)}. \tag{2.6}$$

The cutting rate should increase as $P = F/\pi r^2$ approaches P_0 and should then remain constant. The rate should then remain constant. The rate should be linearly related to the grain size, the frequency, and the amplitude; the relation of rate to pressure agrees with Goetze's results [15-17].

This theory is not accurate, although it agrees with observed relationships in some respects. The form of the basis equation has no logical justification; for instance, it is baldly assumed that the rate is dependent on the work of hardening. The volume cut is dependent on the amplitude as well as on the force, but in Miller's treatment the amplitude affects the rate only via the number of particles under the tool. No allowance is made for the variation in grain size; in fact, all grains are assumed to be cubes of side d. This led Miller to deduce (incorrectly) that all grains take part in the cutting, and he deduced the number of grains on the assumption that the suspension is drawn in when the tool recedes. Neppiras has pointed out that such a mechanism can operate only at low frequencies. Miller considered plastic materials, whereas most materials that are worked in this way are brittle, so the rate cannot be dependent on plastic deformation but must be governed by the size of the particles cut.

Shaw [19] assumed that two processes were involved: direct impact of the tool on grains in contact with the workpiece and impact of moving grains. In both cases v is proportional to the volume V of material dislodged per impact, to N (the number of particles making impacts per cycle), and to f:

$$v \approx VNf. \tag{2.7}$$

Show supposed that the particles were identical spheres of diameter d equal to the mean grain size; then the diameter D of an indent is

$$D = 2\sqrt{dh},$$

in which h is the depth; the volume removed is $V \approx D^3$. Then

$$V \approx (dh)^{3/2} Nf. \tag{2.7a}$$

The depth of the indent may be found as follows. The mean speed of tool or workpiece is low, so the mean static force F applied to the tool must equal the mean force of impact of the tool on the grains:

$$F = \frac{1}{T} \int_0^T F\,(t)\,dt. \tag{2.9}$$

Further, the momentum from a blow is roughly

$$\int_0^T F\,(t)\,dt \simeq \frac{F_{\max}}{2}\,\Delta t. \tag{2.8}$$

The tool moves from its mean position towards the workpiece and at time $t = T/4 - \Delta t$ strikes the grains lying on the surface; these grains are driven into the surfaces during an interval $T/4 - \Delta t$ to $T/4$ to depths of h_w and h_t, respectively; the amplitude of the displacement is ξ_m. The duration of the process is

$$\Delta t = \frac{h_w + h_t}{\xi_m}\,\frac{T}{4}. \tag{2.10}$$

Then we have

$$F_{\max} = \frac{4F\xi_m}{h_w + h_t}. \tag{2.11}$$

The maximum stress σ corresponding to penetration to a depth h is F_{max}/N (force per particle) divided by the area:

$$\sigma_w = \frac{F_{max}}{\pi N d h_w} = \frac{4F\xi_m}{\pi N d h_w \left(h_w + h_t\right)}. \qquad (2.12)$$

Here σ_w corresponds to H, the Brinell hardness [20]. The penetration into the tool is similar. The ratio of the depths is inversely proportional to the ratio of the stresses, which is denoted by q:

$$\frac{h_t}{h_w} = \frac{\sigma_w}{\sigma_t} = q. \qquad (2.13)$$

Then

$$h_w^2 = \frac{4F\xi_m}{\pi N d H \left(1 + q\right)}. \qquad (2.14)$$

The number of particles is inversely proportional to the square of the diameter of each for a tool of fixed area:

$$N = \frac{\varkappa C}{d^2},$$

in which $\varkappa$ is a coefficient of proportionality and C is the concentration; then the depth of the impression in the workpiece is

$$h_w = \left[\frac{4\xi_m F d}{\pi \varkappa H \left(1 + q\right)}\right]^{1/2}. \qquad (2.15)$$

Shaw also deduced the depth of the impression produced by a freely moving particle; here collision with the end of the tool gives the particle a maximum forward velocity $v_m = 2\pi f \xi_m$. The kinetic energy of a particle is

$$E = \frac{1}{2}\left(\pi d^3 \rho\right) 4\pi^2 f^2 \xi_m^2. \qquad (2.16)$$

The particle enters the surface to a depth h'_w, thereby doing work $A = E = Fh'_w/2$. Now $H = F/\pi d h'_w$, so we have

$$h_w = \pi \sqrt{\frac{2}{3}\frac{\rho}{H}} d f \xi_m. \qquad (2.17)$$

For $\xi_m = 75\,\mu$, f = 25 kc, F = 4.3 kg, d = 3.5μ (800 mesh), and C = 0.35 (35% by volume) the result for glass is only 3% of the rate of removal. The vast bulk of the material is therefore removed by directly struck particles.

Substituting for h_w and N from (2.15) and (2.14) in (2.7a), we have

$$v \approx \frac{\xi_m^{3/4} f d^{1/4} F^{3/4}}{H^{3/4}}. \qquad (2.18)$$

The grains have been assumed spherical, which gives $v \propto d^{1/4}$, whereas the actual rate is proportional to the grain size. Shaw explained this discrepancy by assuming that projections on the grains of diameter $d_1 \propto d^2$ are responsible for the effect when the depth of penetration is small. Then a similar calculation gives the rate as directly proportional to grain size and frequency:

$$v \propto f d \left(\xi_m F\right)^{3/4}. \qquad (2.19)$$

This theory is based on a correct conception of the process, for the formation of chips at the instant of impact has been confirmed by subsequent experiments (section 9). All the same, Shaw did not obtain the correct result for the effects of frequency, amplitude, and force. In particular, the particles are very much less regular than he supposed. A few large particles stand out above the others, so only these are effective. The theory also implies that the rate increases indefinitely with the force, whereas in fact there is an optimum force; the fall at high loads is caused by crushing of the grains, which is not reflected in the theory. Further, no allowance is made for the replacement of crushed grains, although this has a marked effect on the rate.

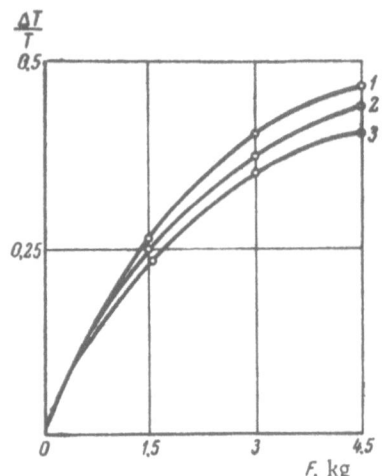

Fig. 29. Time of contact as a function of force for various amplitudes ξ_m (μ) as follows: 1) 15; 2) 30; 3) 45 [21].

For this reason Dikushin and Barke [21, 22] tried to relate the energy consumed to the amplitude and force on the basis of the laws of conservation of energy and momentum.

The mean force when a body strikes an obstacle is

$$F = \frac{Mv\,(1+k)}{\Delta T}, \qquad (2.20)$$

in which M and v are the mass and velocity before the impact and k (the coefficient of restitution) is a constant, being governed by the properties of body and obstacle; ΔT is the time of contact. The change in kinetic energy in this time and the power per blow are respectively

$$\Delta E = E_2 - E_1 = -\frac{Mv^2}{2}\,(1-k^2) \qquad (2.21)$$

and

$$\Delta W = \frac{\Delta E}{T} = \frac{vF}{2}\,(1-k^2). \qquad (2.22)$$

The part of the concentrator (together with the tool) from the end as far as the first displacement node may be treated as a mass vibrating in response to an external force; the motion is sinusoidal up to the time of impact. Here v and F are dependent on the moment of impact. On the assumption that the points of contact and separation are symmetrically disposed with respect to the maximum displacement, we have

$$v_k = v_m \sin \pi \left(\frac{\Delta T}{T}\right), \qquad (2.23)$$

$$F = \frac{Mv\,(1+k)}{\Delta T} = \frac{\pi M\,(1+k)}{T}\,v_m\,\frac{\sin \pi \left(\frac{\Delta T}{T}\right)}{\pi \left(\frac{\Delta T}{T}\right)} = F_0\,\frac{\sin \pi \left(\frac{\Delta T}{T}\right)}{\pi \left(\frac{\Delta T}{T}\right)}. \qquad (2.24)$$

Substituting for the velocity and force in (2.22) we have

$$\Delta W = -\frac{vF_0}{2}\,(1-k^2)\,\frac{\sin^2 \pi \left(\frac{\Delta T}{T}\right)}{\pi \left(\frac{\Delta T}{T}\right)} = \pi\,\frac{v_m^2 M}{2T}\,(1-k^2)\,\frac{\sin^2 \pi \left(\frac{\Delta T}{T}\right)}{\pi \left(\frac{\Delta T}{T}\right)}. \qquad (2.25)$$

This shows that the power is proportional to the above mass and to the square of the amplitude; it is also dependent on $\Delta T/T$ and is maximal when this is 0.37. Tests showed that the time of contact is governed largely by the external force and does not vary much with the amplitude. Measurements were made of the time of contact between the tool and a steel plate for three amplitudes and forces (Fig. 29). The velocity of oscillation is proportional to the amplitude, so the relative power per blow is

$$\frac{\Delta W}{\Delta W_{max}} = \frac{\xi_m^2}{\xi_{m\,max}^2}\,\frac{\sin^2 \pi \left(\frac{\Delta T}{T}\right)}{\sin^2 \pi \left(\frac{\Delta T}{T}\right)_{opt}} \times \frac{\left(\frac{\Delta T}{T}\right)_{opt}}{\left(\frac{\Delta T}{T}\right)}. \qquad (2.26)$$

The relative power was also found as a function of amplitude and pressure:

$$\frac{\Delta W}{\Delta W_{max}} = f\left(\frac{\xi_m}{\xi_{m\,max}}, \quad \frac{P}{P_{max}}\right)\,\xi_{m\,max} = 45\ \mu, \quad P_{max} = 4.5\ \text{kg/cm}^2$$

30

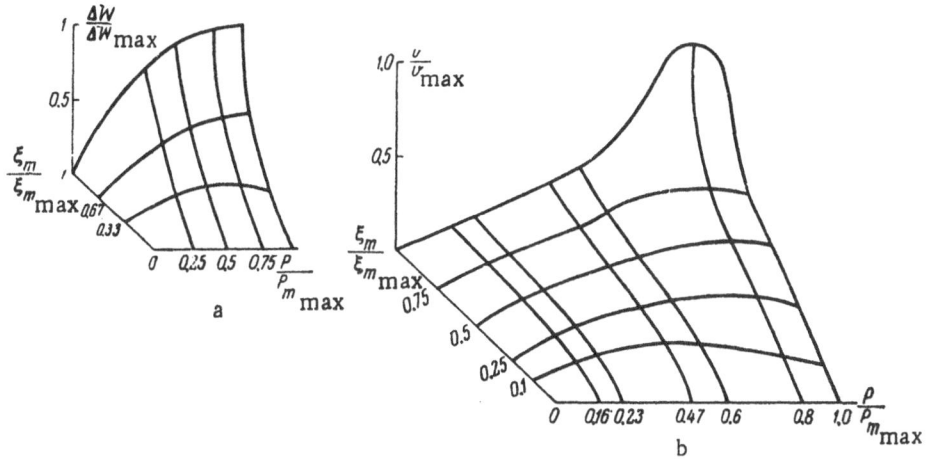

Fig. 30. a) Theoretical relation of power per impact to amplitude and force [22], b) measured cutting rate as a function of amplitude and force.

The $\Delta T/T$ for each P/P_{max} were deduced from the previous experiment; Figure 30a shows the result, which was compared with the calculated relation for the relative rate $\dfrac{v}{v_{max}} = \varphi\left(\dfrac{\xi_m}{\xi_m},\ \dfrac{P}{P_{max}}\right)$. A glass plate was ground for 5 min with M10 corundum powder (40-50%). Figure 30b shows the relative cutting rate, amplitude, and pressure ($\xi_m = 45\,\mu$, $P_{max} = 4$ kg/cm^2, $v_{max} = 10$ mm^3/min).

The results agree well with the theory. The experimental maximum rate appears a little earlier than the theory predicts, which was explained by a fall in the amplitude as the pressure increased.

This comparison indicates that the rate of removal is roughly proportional to the power per impact, so the impacts cause the cutting; as these occur via the abrasive, the damage is caused either via direct impact on the grains or via hydraulic shock. The latter may cause the suspension to enter the cracks and so may affect the rate.

These results are of value as regards technical cutting, but they cannot give a full picture of the whole process. The physics of the process long remained unclear.

8. High-Speed Cinematography

Several lines of research on the physics of the process have been pursued at the Acoustics Institute of the Academy of Sciences. High-speed cinematography was used, in view of the numerous factors that influence the result* [24, 25]. Difficulties arose over the observation of small grains in an opaque slurry between the tool and the workpiece; Figure 31 shows the system used to overcome these difficulties. A glass plate was clamped between two thicker glass plates, the edge of this forming the bottom of a rectangular cell, the other plates forming the sides. The tool (a thin steel blade) was inserted in this cell. The suspension filled the gap between the tool and the plate; the grains could then be seen individually at right angles to the cell.

An FP-22 camera (8 mm film) was used at speeds of 20,000 and 50,000 frames per second. The high-power light source gave sufficient contrast at high speeds, while the lens had a magnification of 1.6 to give a field of view 2.2 × 3 mm.

A frequency of only 6.8 kc was used, in order to increase the number of frames per cycle. The weight of the vibrator (160 g) was sufficient loading. The mean grain size was 220 μ (range 150 to 440 μ); the concentration was usually 50% by weight. Table 11 gives the amplitudes and other basic parameters, which correspond generally to those used in actual machining.

---

* The work was done in collaboration with S. R. Zhukovskii of the Department of Scientific Photography at Moscow University.

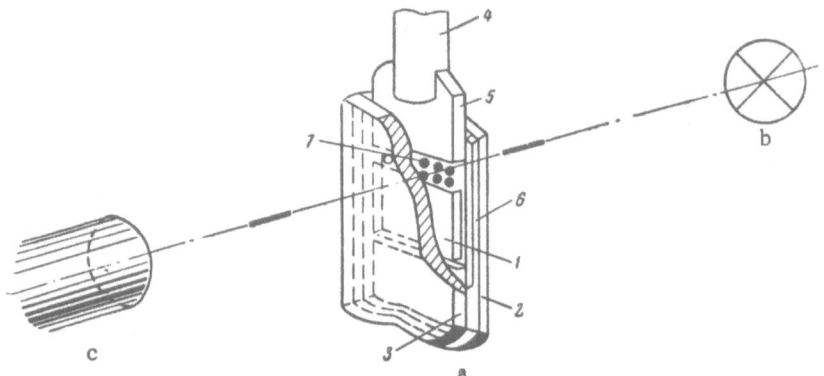

Fig. 31. The cell system: a) cell; b) light source; c) lens; 1) glass workpiece; 2) and 3) glass sides; 4) waveguide; 5) tool; 6) layer of cement; 7) particles of abrasive in water.

There are several views on the nature of the forces. Some explain the erosion as occurring via chipping when the tool strikes the settled particles [9, 19]; others assume that the tool strikes suspended particles. Some [26-28] think that collapsing cavitation bubbles produce the effect either at the particles or even on the surface. These ideas were tested by suspending the tool so that in some cases it did not rest on the particles.

The general nature of the process became apparent on viewing the film at 24 frames per sec. The glass was cut only when the tool struck particles resting on it* ; the damage took three forms. Most of the pits were much smaller than the particles, which were struck periodically by the tool. Cracks and small pits were thereby formed. Figure 32a shows the boundary before the experiment started (film 80); the light part at the bottom is the glass plate, and the dark part at the top is the tool. The abrasive can be seen suspended in the water. The surface of the glass is reasonably flat. The second frame (Fig. 32b) shows the same area 15 msec after the start; the initially smooth boundary has acquired some pits.

In some cases the glass cut can be seen as a finely divided powder, as on the right in the above frame. Repeated blows on a particle of diameter $430\,\mu$ (much larger than the others) resulted in a pit $35\,\mu$ deep.

The third form of damage is relatively rare; as in the previous case, it is associated with exceptionally large particles. The tool strikes a large particle to the right of several small ones (Fig. 33a), which gives rise to a crack (growth time less than 0.02 msec). The fragment of glass corresponds in size to the grain of abrasive. This type of damage was observed in later experiments, in which a slot was cut in a solid piece of glass. It occurs when the number of grains under the tool is small, for then it becomes more probable that there will be one grain much larger than the others.

Other causes of grain movement do not result in damage; in particular, no damage was ever seen from the tool striking a suspended grain, although the speeds thereby attained ranged up to 190 cm /sec. This agrees

TABLE 11. Conditions Used with High-Speed Cinematography

Film	Frames per sec	Position of tool	Gap, mm	Liquid	Ampl., μ
80	20,000	Lying on abrasive	0.34		56
85	20,000	The same	0.29		59
88	50,000	The same	0.25	Water	51
83	20,000	Suspended	0.37		63
86	20,000	The same	0.75		73
84	20,000	The same	0.83	Glycerol	

* Similar conclusions were drawn later [29] by the same method.

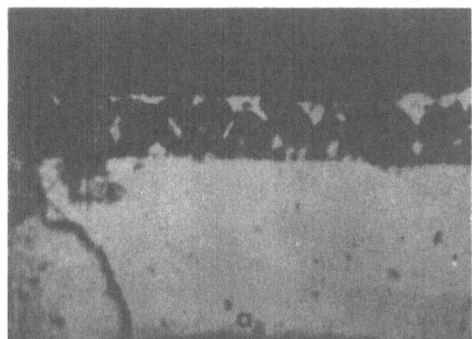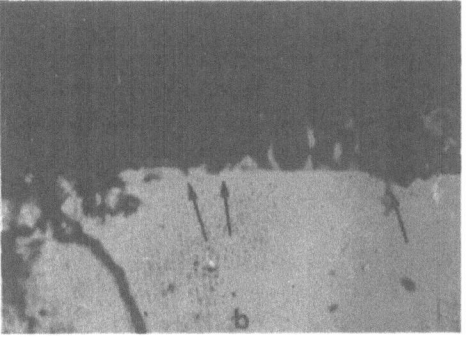

Fig. 32. Surface of glass a) before start and b) 15 msec after start. ×10.

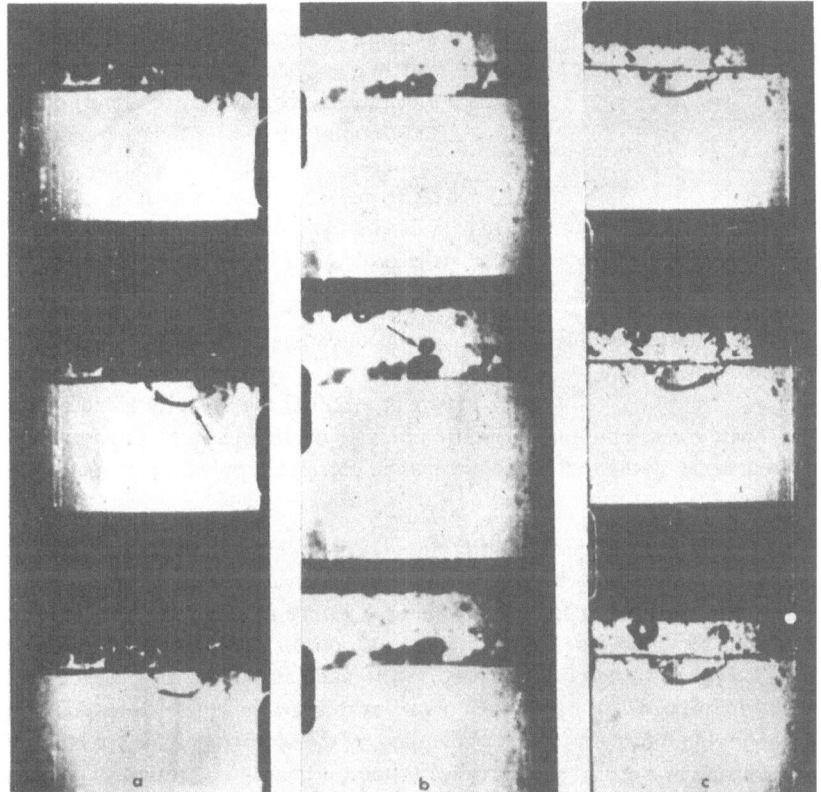

Fig. 33. Frames taken during cutting (×10): a) formation of large pit; b) collapse of cavitation bubble near a particle; c) damage to surface of tool from vibration of a cavitation bubble.

with Shaw's [19] finding that free grains produce very little cutting. Further, the cavitation theory is entirely rejected; Figure 33b shows a bubble directly above a particle lying on the glass, but its collapse produced no signs of damage.

Cavitation does damage the tool to some extent; this is made of softer material. Purely cavitational erosion of the tool was seen on one film; a bubble formed and collapsed repeatedly for many cycles. Figure 33c shows the phases of the bubble for one cycle.

Figure 34 shows the radius as a function of time; the motion is clearly unsymmetrical, and the collapse is extremely rapid. The resulting shock waves damage the material. Figure 33c shows the depression in the

TABLE 12. Ratio of Wear Rates for Vibrating and Fixed Parts
Made of the Same Material

Material	Thickness, mm	Ratio of tool wear rate to depth of hole
Permendur	0.1	1.14
Ditto	0.2	1.24
K-65 alloy	0.2	1.42
U-10 steel (quenched)	0.25	1.20
Ditto	0.50	1.27
Copper	0.5	1.38

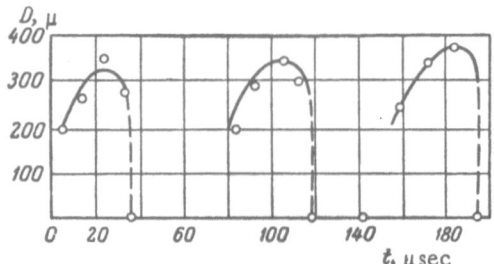

Fig. 34. Diameter of the cavitation bubble of Fig. 33 as a function of time.

surface of the tool. A rough calculation shows that the damage was considerable because the tool was made of brittle quenched steel and was thin. These localized bubbles produce characteristic craters on the tool. Cavitation most commonly occurs near moving surfaces, so the tool is more liable to be damaged, as has been found [10] in other experiments with tools of the same material. The wear on the vibrating part is always greater than that on the fixed one (Table 12).

The bubbles do not exert sufficient force on the grains to damage the surface. Some frames show grains set in motion by shock waves from bubbles of diameter about 0.1 mm; the velocity is measured as 30-40 cm/sec.

On the other hand, we should not assume that cavitation plays a negligible part; the same applies to the continuous flow of liquid generated by the tool, which keeps the abrasive moving. This motion is essential to the continuance of the cutting.

The cutting rate falls if more viscous liquids are used. Table 8 gives the cutting rate in water-glycerol mixtures; it changes by a factor of 100 when the glycerol content varies from 0 to 100%. Glycerol gave more cavitation bubbles than water did even at lower amplitudes; the flow rates for glycerol were much lower. Figure 35a shows the motion of abrasive particles in glycerol (positions at intervals of 2.5 msec, speed about 4 cm/sec); Figure 35b shows glass in water (25 cm/sec). The numbers denote the positions at intervals of 0.5 msec. These are related to the cutting rates in such a way as to show that these flows play a large part in the cutting. On the other hand, the cutting rate is not governed by the viscosity alone, for cavitation also affects the rate; it accelerates cutting by keeping the particles in motion, but it also retards it by tending to displace the abrasive from the working gap. The cavitation and the flow under the tool both affect the mixing, removal, and influx of abrasive.

Ultrasonic cutting involves two essentially different processes: chipping by impact of the tool on the abrasive particles and migration of particles in the working gap. The cutting rate, accuracy, surface finish, and tool wear are governed to various extents by both processes.

9. Nature of the Damage to the Material

We have seen that the material is eroded when the tool strikes the abrasive particles. The maximum speed of the tool is a few meters per second, in which range the hardness of a metal remains nearly constant. This is also the case for brittle materials, for Hooke's law applies almost up to the point of fracture. The stress in an elastic body propagates with the speed of sound; cracks appear when a limiting stress is reached. The speed of propagation for cracks is comparable with the speed of sound (about 1400 m/sec [30]). The size of the chips is independent of the time of contact and of the frequency, for the time taken for a crack to penetrate to the maximum depth (about 10^{-2} cm) is only 10^{-7} sec. This indicates that the fracture of a brittle material may be of quasistatic type; the cutting rate is then independent of the average force and is governed solely by the

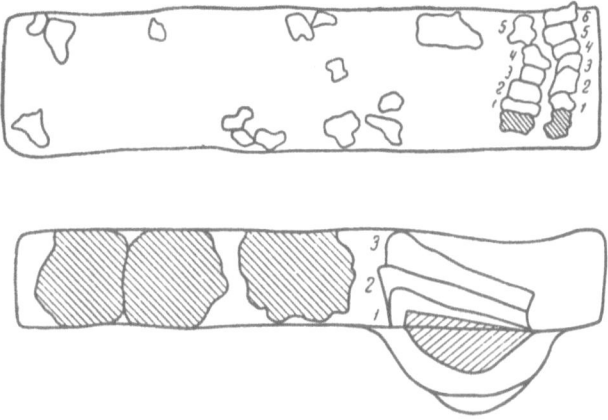

Fig. 35. Motion of abrasive particles in glycerol (film 84)
and of particles of glass in water (film 88).

maximum stress set up by the blows. The volume of damaged material is dependent on the maximal stress and on the grain size. This means that inhomogeneity in the grain size has a marked effect on the damage. The process appears to be as follows. The tool moves up from its most remote position and at some point reaches the largest grains, which are forced into the tool and workpiece. The number of grains and the force they exert increase as the motion continues; some of the grains are fractured. Eventually the tool comes to rest, its distance from the surface corresponding to a certain maximum force. The calculation scheme is then as follows. Let there be N particles between the tool and the surface (Fig. 36), the height ξ being a random quantity. The distribution is a function $\varphi(\xi)$.   At first the tool touches the largest grains (height ξ_m). The process continues as above; we assume that the particles are incompressible, which means that the depth of penetration is $h = \xi - x$, in which x is the mean distance between surface and tool. We assume that we know the relation between the force of Φ_ξ on a particle and h in the form $\Phi_\xi = \Phi_\xi(\xi - x)$ for $\xi - x \geq 0$. Then the total force F acting on the particles is

$$ F = \int_x^{\xi_{cr}} \Phi_\xi(\xi - x) \cdot N\varphi(\xi)\, d\xi \ . \tag{2.27}$$

The particles are of finite strength, so the largest must be broken. This means that the upper limit of integration ξ_{cr} is less than ξ_m.

In the same way we can deduce the volume of material cut per cycle. For this we need to know the relation of the damage produced by a particle to the force acting on it, or, which amounts to the same, to the depth of penetration: $V_\xi(\xi - x) = V_\xi[\Phi_\xi(\xi - x)]$. Then the volume cut per cycle is

$$ V = \int_x^{\xi_m} V_\xi(\xi - x) \cdot N\varphi(\xi) d\xi \ . \tag{2.28}$$

It is also readily shown that the number of chips is

$$ n = \int_x^{\xi_m} N\varphi(\xi) d\xi \ , \tag{2.29}$$

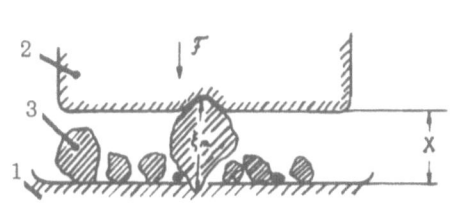

Fig. 36. Scheme for ultrasonic cutting:
1) workpiece surface; 2) tool surface; 3) particles of abrasive.

and that the mean depth of penetration is

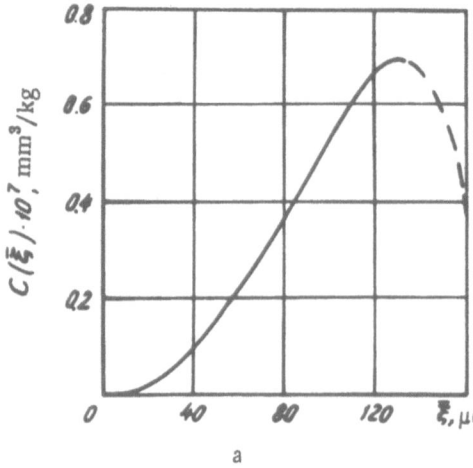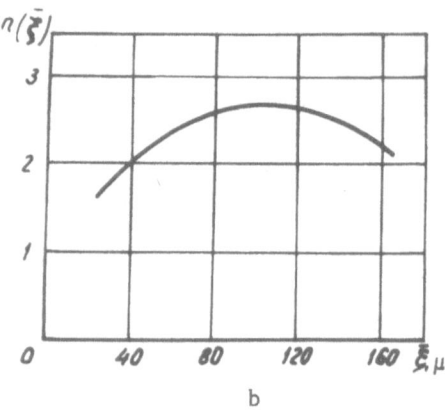

a   b

Fig. 37. Relation of C and n of (2.34) to mean grain size.

$$\overline{h} = \frac{\int_{x}^{\xi_m} H_{\xi}(\xi - x) \cdot N\varphi(\xi)d\xi}{n} \qquad (2.30)$$

in which $H_{\xi}(\xi - x) = H_{\xi}(\Phi_{\xi})$ is the relation of depth to force for a single particle.

The experimental evidence has enabled us to deduce the particle size distribution for an abrasive of any grain size:

$$\varphi(\xi) = 1.095 \frac{N}{\overline{\xi}} \left[1 - \left(\frac{\xi}{\overline{\xi}} - 1 \right)^2 \right]^3 \qquad (2.31)$$

in which $\overline{\xi}$ is the mean size of the particles in the working gap. On the assumption that the particles may be treated as spheres having projections whose radii of curvature are proportional to the mean dimension of the particle, we have the relation of $h = \xi - x$ and Φ; this relation has been deduced from the literature on the damage resulting from indentation with abrasive grains (see [51, 52], for example). This takes the form

$$\Phi_{\xi}(\xi - x) = \begin{cases} 4H_t \overline{\xi}(\xi - x) & \text{for } \xi - x \le l_1 \\ \dfrac{4H_t \overline{\xi}}{1 + q_1 \overline{\xi}} (\xi - x - q_1 l_1) & \text{for } l_1 \le \xi - x \le l_2 \\ 0 & \text{for } \xi - x \ge l_2 \end{cases} \qquad (2.32)$$

in which $l_1 = b_0 y/4H_t$ and $l_2 = 1/q_3 + (q_1/q_3)\overline{\xi} - q_1 l_1 \overline{\xi}$.

The notation is as follows: $q_1 = 4H_t/a_0 H_w$, $q_3 = 4H_t/c_0 H_a$; a_0, b_0, and c_0 are not dependent on the properties of the tool, workpiece, or abrasive, whose hardnesses are respectively H_t, H_w, and H_a. The parameter y is the ratio of the diameter of a projection to the mean diameter of the grain. We have also been able to relate the volume damaged to $h = \xi - x$:

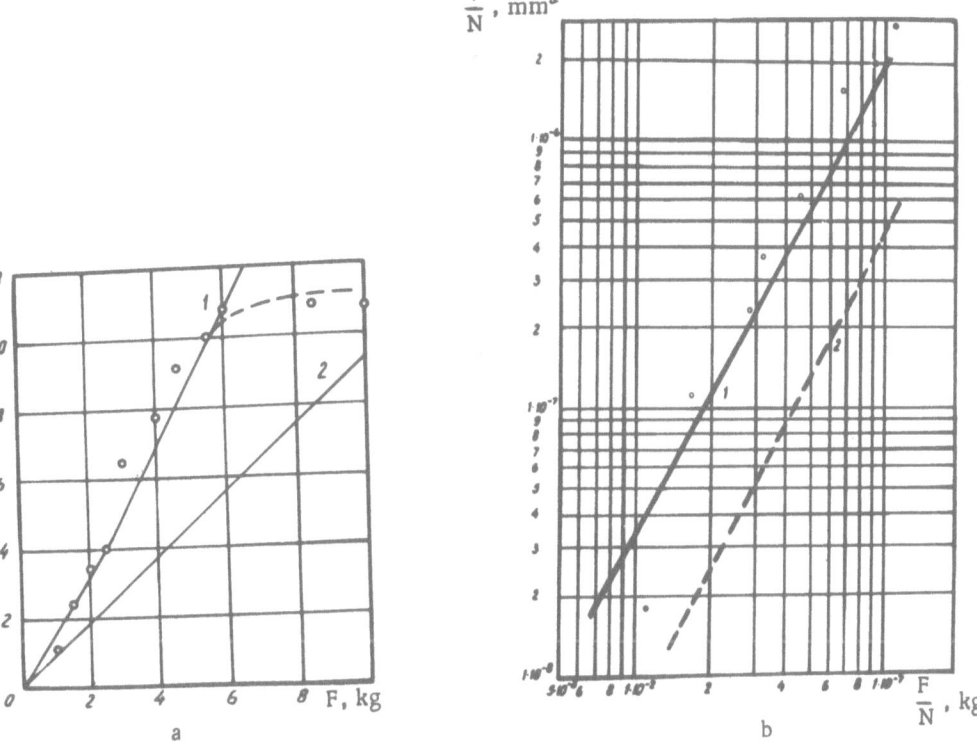

Fig. 38. Relation to force of a) number of chips and b) volume removed. The straight lines correspond to the following y: 1) 0.1; 2) 0.6.

$$V_{\xi}(\xi - x) = \begin{cases} 0 & \text{for } \xi - x \le l_1 \\ \dfrac{K_0 q_1^3 \overline{\xi^3}}{(1 + q_1 \overline{\xi})^3} \, (\xi - x + l_1)^3 & \text{for } l_1 \le \xi - x \le l_2 \\ \dfrac{K_0 q_1^3 \overline{\xi^3}}{q_3^3} \, (1 - q_3 l_1)^3 & \text{for } \xi - x \ge l_2 \end{cases}$$

(2.33)

Here K_0 is a constant. These functions are inserted into (2.27)-(2.29); numerical integration then gives the number of chips as proportional to the mean load. Further, it has been found that the damaged volume for glass worked with boron carbide of various grain sizes is related to the force as follows:

$$\frac{V}{N} = C(\overline{\xi})\left(\frac{F}{N}\right)^{n(\overline{\xi})}$$

(2.34)

Figure 37 shows $C(\overline{\xi})$ and $n(\overline{\xi})$. The coefficient of proportionality is maximal under these conditions for a mean grain size of $130\,\mu$; the maximum exponent is roughly three for a mean grain size of about $100\,\mu$.

The theoretical relation has been tested by experiment.[*] The carbide (80 mesh) was poured as a layer onto the surface of a glass plate; then a rod 7 mm in diameter was slowly pressed onto the layer with its end strictly parallel to the surface. The force and the number of particles under the rod varied from test to test. The number and size of the impressions were established by examination under the microscope.

---

[*] The experiments were performed with the aid of Yu. L. Tisenbaum.

The surface concentration c was kept constant at 300 cm^{-2} in order to find the relation of mean number of indents to force (Fig. 38a); the relation is clearly nearly linear up to 5 kg, but beyond that there is little further advance between 6 and 9 kg. The maximum number of particles giving marks did not exceed 10% of the total under the end of the tool. The results show that the general trend is in accordance with the prediction from (2.32); the angularity of the particles, as determined by direct observation, was about 0.1. The relation deviates from linear above 6 kg, on account of marked increase in the area of contact; the damage is then produced by the whole grain rather than by the separate projections.

Each projection may be represented as a cone in order to deduce the volume of material so cut. Figure 38b shows experimental results together with values calculated from (2.27), (2.28), (2.31), (2.32), and (2.33). The agreement is clearly best for y = 0.1.

We have analyzed Ohira's results [53] by the above method; they also gave agreement with the theory.

This agreement confirms that the size distribution of the particles has a marked effect on the damage. Further, it shows that the volume cut is related to the stress by a simple power law.

The damage is governed solely by the force acting at the given moment, so the volume cut by a single blow corresponds to the maximal force F_m and is as though that force were to act continuously. In fact, the volume is given by the formula if in this we substitute the maximal force. The theory, especially this feature of it, has been confirmed in tests on impact damage [32] and also directly by ultrasonic cutting itself.

Nishimura and Shimakawa [32] have shown that the volume cut increases as the duration of the blow decreases if the momentum transfer is constant. They related the volume cut or the cutting rate to the mean force, but we consider it physically sounder to correlate the volume with the maximum force, because the damage is governed by the maximum stress. Their results [32] show that $v = cF_m^3$.

Similar conclusions have been drawn from experiments on the damage to the material directly during ultrasonic cutting [31] by means of high-speed cinematography applied to a 4770 machine (250 W, about 18 kc).* The tool was a plate 0.1 mm thick and 37 mm wide soldered to the end of a stepped concentrator driven with an amplitude of 24 μ. This tool was used to cut a slot in a block of glass, the plane of the slot being parallel to the long side face. The abrasive was 120 mesh boron carbide suspended in water. The suspension was retained by a cell cemented to the upper face.

The process was recorded with an SKS-1 camera with the axis perpendicular to the plane of the cut; a Jupiter-12 objective (35 mm) was used to give a magnification of 35. One frame then represented a field 0.22 × 0.3 mm. A mercury arc (SVDSh-250) with condenser lenses provided sufficient intensity for use in transmission.

The concentration was high, and the side faces were rough, so the motion of the individual particles could not be seen at speeds of 1500 to 5500 frames per second; but the effects at the boundary were clearly followed.

Most of the material is removed by almost instantaneous formation of individual chips; no correlation between the sites of damage could be observed. Miller and Shaw's assumption (simultaneous chipping by all particles) is not confirmed.

We have seen in section 8 that large semicircular pits (depth up to 50μ) are comparatively rare; most are triangular and less deep (not over 10μ). Figure 39 shows such a pit, with several grains of abrasive shown on the same scale. Pits of any size appear in less than the time between frames even at 5500 frames per second.

Figure 40 gives n(τ) = ν exp (−$\nu\tau$), in which τ is the time between production of adjacent pits. This distribution is characteristic of random sequences of rare events, in which $\nu = 1/\bar{\tau}$ is the mean number of pits per unit time, $\bar{\tau}$ being the mean interval between events.

Under the conditions used, 20-30 pits were produced per second. From the ratio of length to width it is clear that one pit is produced per several cycles over the working area; there were some dozens of grains in the

---

* S. I. Glazkov analyzed the results.

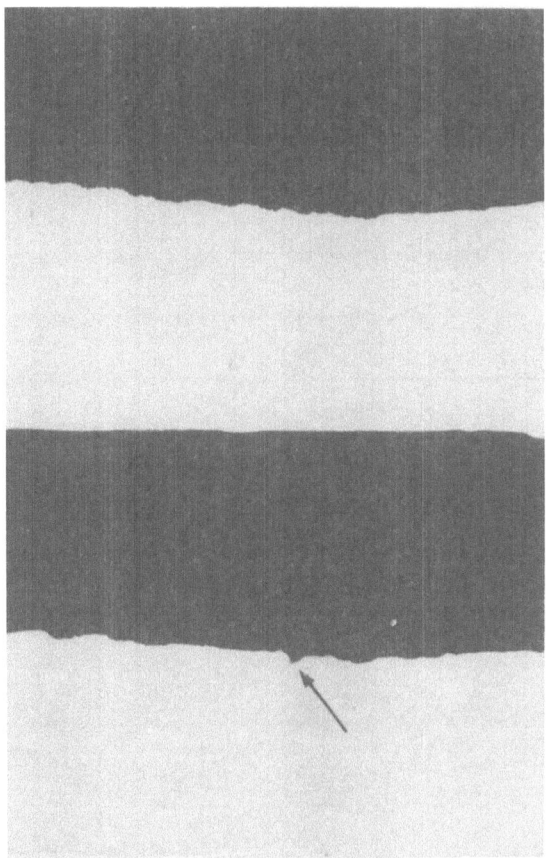

Fig. 39. Formation of scratch on surface of workpiece
(×200).

working gap (mean diameter about 100μ), so the probability of production of a pit by a particle in one cycle ranges from 0.001 to 0.01. This very low probability is not characteristic of thin cuts alone. Indirect evidence (as from the size of the cut particles or from the mean height of the roughness) indicates that pit formation is comparatively rare.

The mean number of pits per unit time increases with the force:

F	1.3	1.8	2.4
$\bar{\nu}$	21.6	27.8	32

The maximum force increases with the static force [33], so this result is analogous to the relation of number of cuts to static force. The depth distribution of the pits has been deduced by experiment (Fig. 41); the number rises steadily towards the smaller sizes, as theory predicts.

Formula (2.31) implies that the distribution in height for the largest particles is $N_j \propto (\xi_m - \xi_j)^3$, in which $\xi_m = 2\bar{\xi}$. The pit depth should show the same distribution.

Pit-depth distributions for various static forces show that the maximum depth increases with the static force.

Comparison of theory with experiment shows that the rate of ultrasonic cutting is governed by the distribution in height of the abrasive particles and by the maximum stress; the volume cut is $v \propto F_m^n$, in which F_m is the maximum force. For glass cut with boron carbide of 80-120 mesh, $v \propto F_m^3$.

The force resulting from the blows of the tool cannot be taken as known, for it is itself governed by several parameters, particularly the amplitude and the static pressure.

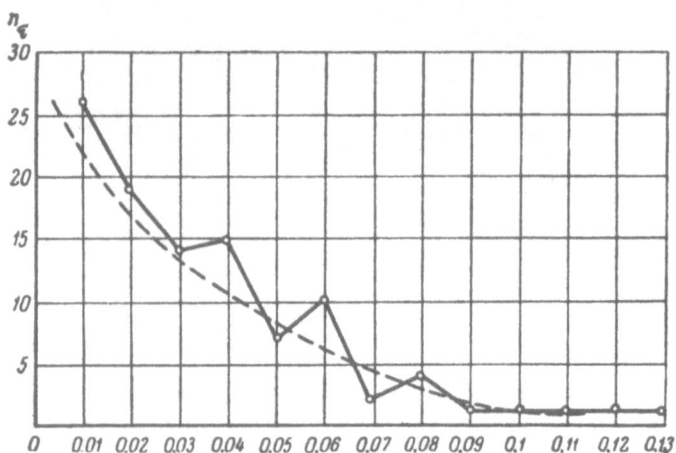

Fig. 40. Relation of n_τ (number of chips) to interval τ between chips. The broken line represents the theoretical relation $n(\tau)$ $= \nu \exp(-\nu\tau)$.

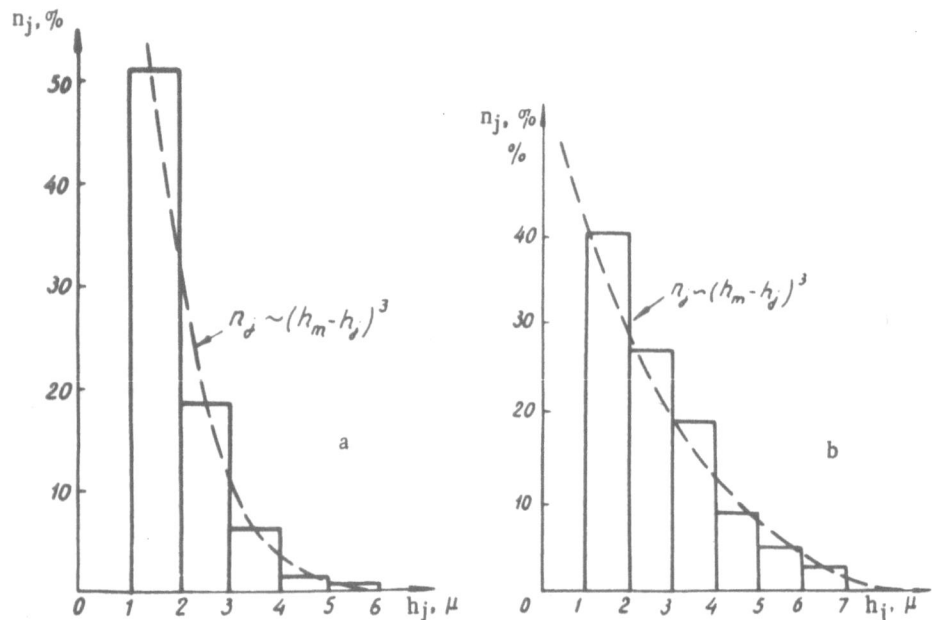

Fig. 41. Depth distribution of pits: a) F = 0.26 kg; b) F = 0.41 kg. The broken line shows the theoretical relation $n_j \propto (h_m - h_j)^3$.

Shaw's theory (section 7) indicates that $F_m \propto (\xi_m F)^{\frac{1}{2}}$. The relation of stress to amplitude and static force has been examined [33-35], a piezoelectric plate being used to measure the forces. A better method [35] is based on the optical anisotropy produced in many transparent bodies (glass in particular) by stress; the intensity transmitted when the specimen is placed between crossed polarizers is related to the instantaneous stress (see [35] for details). Measurements were made of the maximum force (stress); as one would expect, the maximum force was several times the static force (e.g., in one case it was 3.5 kg, when the static force was only 0.6 kg). We have found that the maximum stress is related to amplitude and static pressure by

$$\sigma_m = \varkappa \sqrt[3]{\xi_m^2 p} \ . \tag{2.35}$$

40

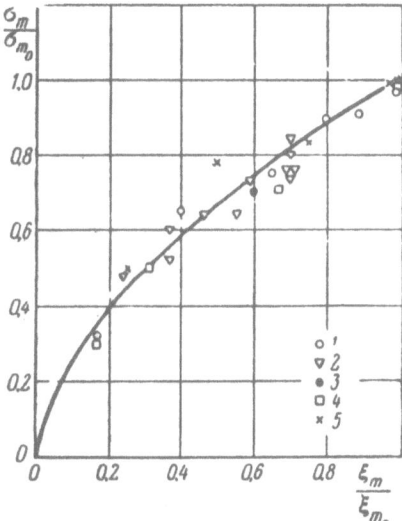

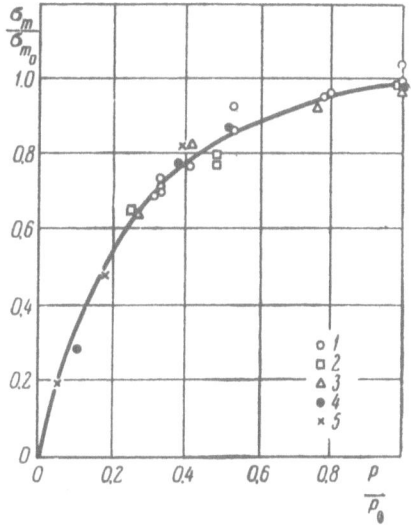

Fig. 42. Relative stress amplitude as a function of relative vibration amplitude for various static pressures p and tool diameters d; 1) 1.6 kg/cm² and 20 mm; 2) 1.6 kg/cm² and 7 mm; 3) 0.3 kg/cm² and 20 mm; 4) 2.4 kg/cm² and 7 mm [44]; 5) 15.7 kg/cm² and 4 mm [33].

Fig. 43. Relative stress amplitude as a function of relative static pressure for various vibration amplitudes ξ_m and tool diameters d: 1) 17 μ and 20 mm; 2) 19.5 μ and 7 mm; 3) 12 μ and 7 mm; 4) 40 μ and 4 mm [33]; 5) 30 μ and 4 mm [33].

This does not agree with Shaw's theory; in particular, the effects of amplitude and pressure on the rate are different. All the published results for various conditions in fact fit this formula.

Figure 42 shows values of σ_m/σ_{m_0} and ξ_m/ξ_{m_0} collected from various sources [33-35], which all fit a curve $\sigma_m/\sigma_{m_0} = (\xi_m/\xi_{m_0})^{2/3}$ no matter what the pressure, area, and method used; the spread is only 6-8%. Figure 43 similarly compares the stress with the pressure [33, 35], which implies that $\sigma_m/\sigma_{m_0} = (p/p_0)^{1/3}$.

The volume cut is proportional to the cube of the maximal stress. Figure 42 may be redrawn (Fig. 44a) as the cube of the relative stress against the relative amplitude, on which we show results calculated from values

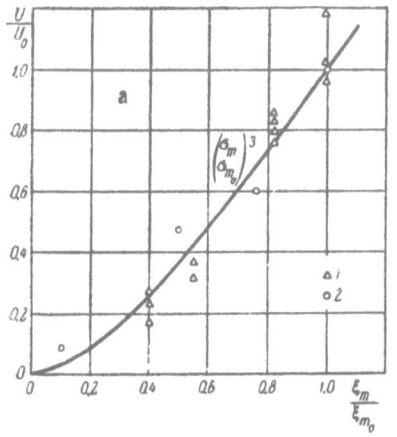

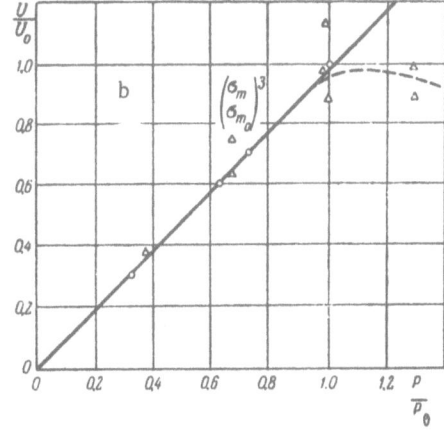

Fig. 44. Relative cutting rate as a function a) of relative amplitude and b) of relative static pressure. The curve shows $(\sigma_m/\sigma_{m_0})^3$ as a function of the same variables.
1) $v_0 = 0.21$ mm/min, $p_0 = 2.2$ kg/cm², $\xi_{m_0} = 11 \mu$ [8]; 2) $v_0 = 0.06$ mm/min, $p_0 = 2.4$ kg/cm², $\xi_{m_0} = 45 \mu$ [21].

given by Neppiras [8] and Dikushin and Barke [21]. The rate as a function of amplitude corresponds to the cube law. Figure 44b shows the same correlation by reference to the relative static pressure; the points lie precisely on a straight line for pressures less than p_0, so the cutting rate and cube of the maximal stress both vary in the same way with the pressure. However, the actual rate is less than the expected one at high pressures.

These results show that $v \propto \xi_m^2 p$ for $p < p_0$.

Increase in amplitude provides the best means of raising the cutting rate, but amplitudes in excess of $50-60\ \mu$ are not readily available. Further increase is provided by a rise in pressure, which increases the cutting efficiency.

10. Cavitation and Motion of the Abrasive Particles in the Working Gap

We have seen in sections 8 and 9 that the material is cut when the tool strikes the grains. The process is random, so a uniform distribution of particles should give rise to conformity between tool and workpiece; but we have seen (section 6) that the recess produced by a flat-ended tool is deeper at the edges. In addition, the surface of the workpiece sometimes shows substantial local defects. These features restrict the use of the method in precision machining.

Veroman [36] has made a detailed study of accuracy and surface finish in the production of hard-alloy dies; he found that pits and grooves are produced on the end of the tool. These pits were produced with hard alloys and, to a smaller extent, with glass. He found that the workpiece surface becomes convex as the tool penetrates, and that the convexity produced with glass was less than that with hard alloys.

A fresh tool with a flat end was used in each case, but convexity always developed, the extent being the greater the longer the time of operation. This means that precision machining demands short times; the tool may also be reground periodically to give a reasonably flat surface at considerable depths. For instance, the deviation from a flat surface could be reduced to 0.1 mm for a hole of depth $1.1 + 0.2 + 0.1 = 1.4$ mm, whereas a single cut to a depth of less than 1.1 mm gave a deviation of 0.22 mm.

Numerous tests showed that the local uneveness is purely random; convexity arises elsewhere if some local convexity is removed. It is considered [34] that the effect is of cavitation origin and is caused by production of cavitation bubbles at the surface of the tool.

Cavitation produces not merely pitting at the surface of the tool but also a change in the distribution (concentration) of the abrasive; the bubbles arise mostly under the center of the tool. It is clear from what has been said above that the change in concentration will affect the cutting rate.

High-speed cinematography enables one to examine the motion of the particles, the role of cavitation, and the mean and local concentrations of abrasive. We have applied it to a flat-ended tool cutting a cylindrical hole in glass* on a 4770 machine. The hole (8 mm diameter) was cut in a plate 7 mm thick with an amplitude of $35\ \mu$ and a load of 1 kg with boron carbide (80 mesh) in water. This coarse grade was used to simplify observation of individual particles. The plate must be horizontal, so the camera was arranged as in Fig. 45. The prism turns the rays through 90°. The SKS-1 camera recorded the picture (ratio 1 : 1) at a maximum speed of 5500 frames per second. The specimen was illuminated at an angle, which eliminated surface reflections. The matt end of the tool provided a uniform light background, against which the dark particles were clearly visible.

Figure 46 shows several frames taken at 4400 frames per second (the broken line shows the end of the tool). The particles at the edge (horizontal arrow) vary in brightness at any given instant; the same applies to the ones at the center (vertical arrows on two frames). These variations in brightness are caused by the irregular shape of the particles, and they make it difficult to follow a single particle from frame to frame.

The results show that the suspension does not completely fill the gap. The concentration is uniform and highest at the start (about 180 particles over the entire surface); after a few seconds, some have been lost from the gap, while the rest lie on a ring whose area is only 2/5 of the total, the concentration here being roughly equal to the initial one.

---

* Similar experiments have been described by Aver'yanova and Milovidov [29].

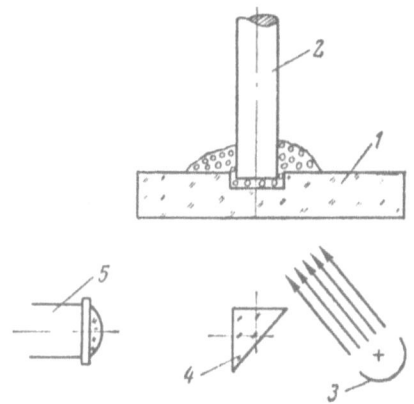

Fig. 45. Side view of system for photographing end of tool: 1) glass plate; 2) tool; 3) SVDSh-250 lamp; 4) glass prism; 5) camera lens.

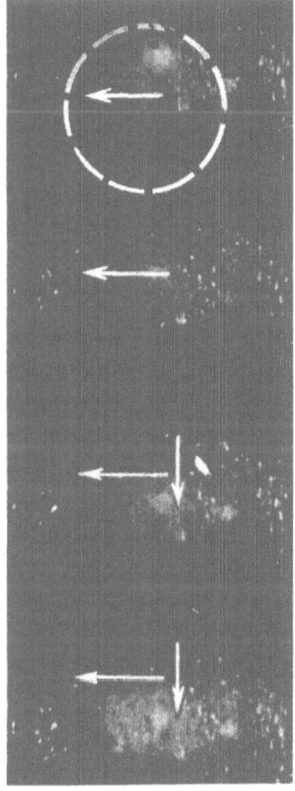

Fig. 46. Frames recorded in high-speed cinematography (×3).

The variation in depth of cut from edge to center reported by Nomoto (Fig. 17) is caused by this uneven distribution, whose main causes can be established by careful examination of the film. The particles at the edge do not move much, but those in the center have speeds of about 0.2 m/sec, so they spend less time there. The cause of this motion is not clear, but it is likely to be the microcurrents [37].

Cavitation has a pronounced effect on the motion; the bubbles appear mainly in the central zone free from abrasive. Some are produced and collapse during a single cycle, while others persist for a long time. Figure 47a shows such bubbles recorded at 5000 frames per second. The frequency was about 18 kc, so the vibration of the bubbles could not be followed. The vigorously pulsating bubbles are in cavitation; a method of using high pressures was proposed in 1957 [38], but Rostovtsev's study of the process was not made till several years later [39, 40]. Figure 48 shows the system. The chamber 4 is connected to a gas cylinder. The concentrator operates via a diaphragm that forms the top seal of the chamber. The amplitude is measured with the microscope 3. Specimens of glass and aluminum were worked with a cylindrical tool 5 mm in diameter at 9.5 and 17 kc with amplitudes up to 9 μ for 2 min (glass) or 3.5 min (aluminum). The force (about 2 kg) is provided by the spring 7. The lower seal is formed by a plate supported by the hydraulic press 5, in order to provide rapid interchange of specimens. The recess in the holder 6 was filled with boron carbide (0.15 g of 180 mesh or 0.14 g of 220 mesh) together with 0.5 cm^3 of the liquid (water, ethanol, salt solution saturated at 15-20°C). The amount cut was determined by weighing.

Figure 49a shows the rate as a function of external pressure for glass at 17 kc with the 180 mesh in water, alcohol, and salt solution. There is a rise to a limiting rate as the pressure is raised; the pressure at which the limit is reached is not dependent on the material or gas and is governed solely by the properties of the liquid. The knee coincides (within the error of experiment) with the amplitude of the acoustic pressure, which is

$$p_m = \rho c \omega \xi_m,$$

in which ρc is the acoustic impedance, ω is the angular frequency, and ξ_m is the amplitude of the vibration.

This formula applies to a traveling plane wave, so it is doubtful whether it applies to the pressure in a narrow gap. The results for the present amplitude and frequency are 9.4 atm (water), 6.3 atm (alcohol), and 13.9 atm (salt solution).

Rostovtsev considers that cavitation ceases when $p_0 = p_m$ if the liquid has not been degassed.

The effects of the hydrostatic pressure are not affected by the frequency and amplitude; the maximum rate is reached when the pressure equals the acoustic pressure, though at 9.5 kc there is some tendency for the rate to fall as the pressure is increased further, especially for alcohol (Fig. 49b). The effect is caused by a

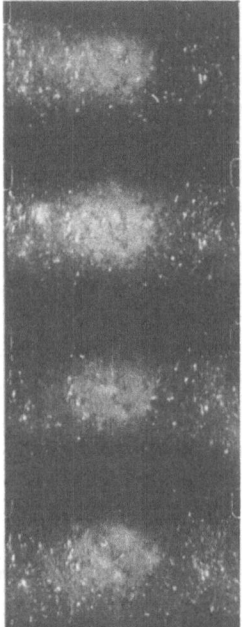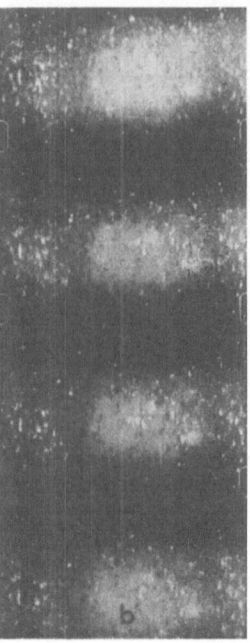

Fig. 47. a) Cavitation in the working gap and b) motion
of particles caused by cavitation.

change in the ratio of the amplitudes at the end of the tool and at the point of measurement, which arises from the changing load on the end of the vibrator. All the same, the rate remains higher than at atmospheric pressure, so suppression of cavitation does accelerate the cutting.

It has been reported [41] that suppression of cavitation reduces the rate for rocks cut at 2.45 kc. A tubular magnetostriction vibrator and the specimen were placed in a pressure chamber; the abrasive (corundum) had a mean grain size of 0.3 mm, and 1 g of it was placed under the tip of the vibrator. The external pressure was varied from 0 to 20 atm by steps of 5 atm. The rate fell by 20% when the pressure was raised by 5 atm; further change was without effect. These tests differed somewhat from those previously described in parameters (frequency, grain size, and so on) and in method; all the abrasive was in the working gap. Further, some parameters of major importance (amplitude, static force) were not stated. Although Rostovtsev's results were more reliable and his methods correspond more closely to those actually used, we need a further study of the effects of cavitation on the rate and also of the effects of high pressures.

The temperature has a marked effect on cavitation, for there is a rapid increase in the number of nuclei as the temperature rises; this is accompanied by a rise in vapor pressure and a fall in the shock on collapse, so the amount of damage per bubble is reduced. The two opposing tendencies result in a maximum in the cavitation damage at a certain temperature. Bebchuk's study of cavitation damage [42] has given us the temperature dependence of cavitation erosion for several liquids. Water has its maximum at 50°C; a change of ±20°C causes a fall of 30%. The maxima for carbon tetrachloride, acetone, alcohol, and benzine lie in the range 0 to 10°C, and their sizes are 3 times less than that for water.

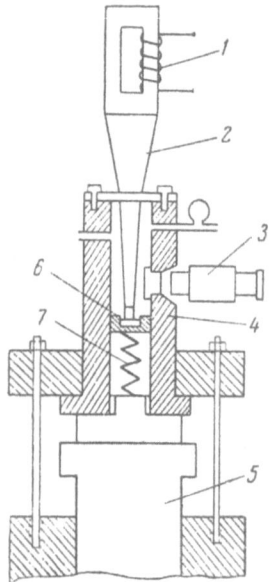

Fig. 48. Apparatus used in experiments at high pressures: 1) magnetostriction source; 2) concentrator; 3) microscope; 4) chamber; 5) press; 6) cell; 7) spring.

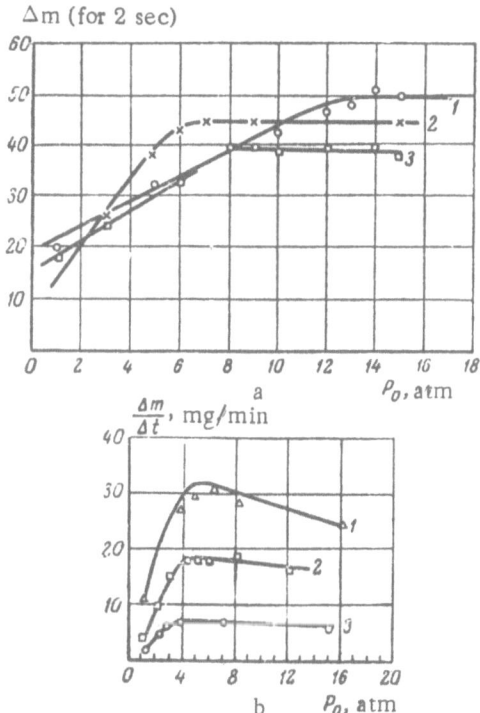

Fig. 49. Cutting rate as a function of external pressure [40]: a) 17 kc; 1) salt solution, 2) alcohol, 3) water; b) alcohol at 9.5 kc and amplitudes ξ_m of 1) 7.6 μ, 2) 6 μ, 3) 4.4 μ.

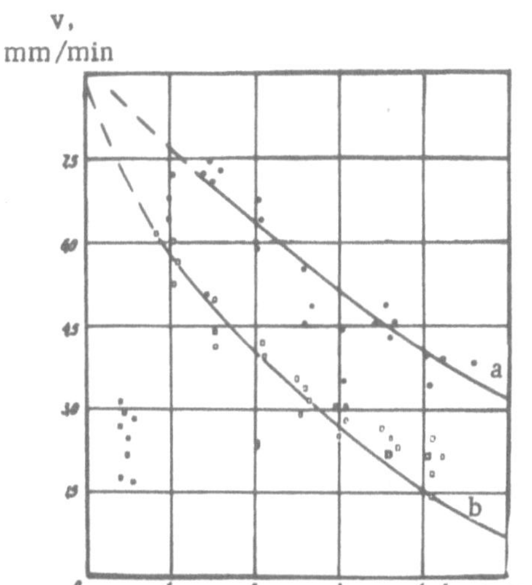

Fig. 50. Rate of cutting as a function of depth at: a) 18°C; b) 60°C.

We [43] have made measurements on the cutting rate as a function of temperature for water over the range 18 to 60°C by means of holes 6 mm in diameter and up to 4 mm deep in glass 6-7 mm thick at constant amplitude and static force. The liquid was heated by passage through a thermostatic jacket. The abrasive was boron carbide (120 mesh). The system took several hours to come to equilibrium, after which the error in the temperature did not exceed ±1°C.

Measurements were made of the time to reach a given depth at intervals of 0.5 mm up to 3 mm total. Figure 50 shows some results. An interesting point is that the first tenth of a millimeter or so is cut at a lower rate at any temperature, but this is not the result of a harder surface layer, for the rate is unaffected if the surface layer is first removed. The rate becomes maximal at a depth of 1.0-1.5 mm, the maximum being nearly independent of temperature. Further increase in depth causes a fall at all temperatures. The slight fall in the maximum rate at the higher temperatures confirms that cavitation has no pronounced effect; results similar to Bebchuk's would have been obtained if this were not so [42].

On the other hand, cavitation erosion of the tool undoubtedly increases when the water temperature is raised to 60°C. Figure 50 shows that the maximum rate varies a little with temperature, the relative fall in the rate at a given depth increasing with the temperature. The change in rate with depth is largely the result of altered flow speed of the suspension in the lateral gap, which is dependent on the relation between cavitation and rate of penetration of the abrasive to the bottom.

11. Relation of Cutting Rate to Depth of Tool

We have seen above that the cutting rate fell beyond depths of 1.0-1.5 mm; this occurred under almost all conditions, but the precise behavior of the rate was dependent on the area and shape of the hole, on the force, on the abrasive, and so on. Two explanations are possible. The first [14] is that the area of the side surface increases with depth, which increases the friction and so reduces the amplitude, which governs the cutting rate. However, experiment does not confirm this. A circular stepped concentrator having its end of length $3\lambda/4$ was used to cut a hole in glass 8 mm thick; the amplitude was unaffected (ξ_m = 35 ± 1 μ) down to depths of 8 mm.

The second is that the depth impairs the inflow of fresh abrasive and removal of broken material. This argument may be tested as follows. We have seen in section 8 that the true concentration of abrasive in the working gap decreases when the depth becomes large; but the main cause of the fall in rate is that the grains are broken up (the grain size has a marked effect on the cutting rate). A high rate of breakage will mean that the cutting rate is governed by the rate of influx of fresh grains, which itself is dependent on the depth of the hole. Measurements should therefore be made of the rates of breakages and influx.

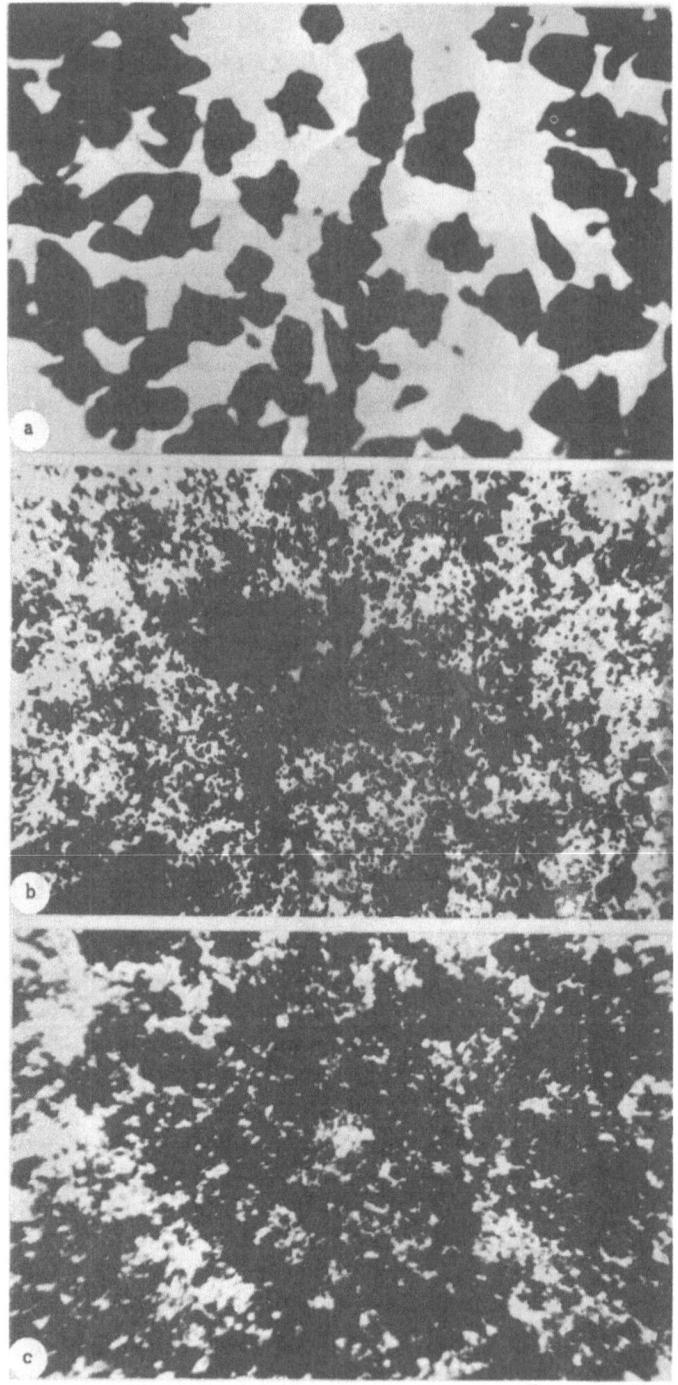

Fig. 51. Boron carbide, 120 mesh: a) before use; b) among small
cut particles; c) after use for 0.5 sec.

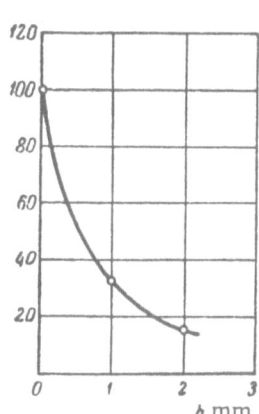

Fig. 52. Relation of grain size
of slurry to depth of hole.

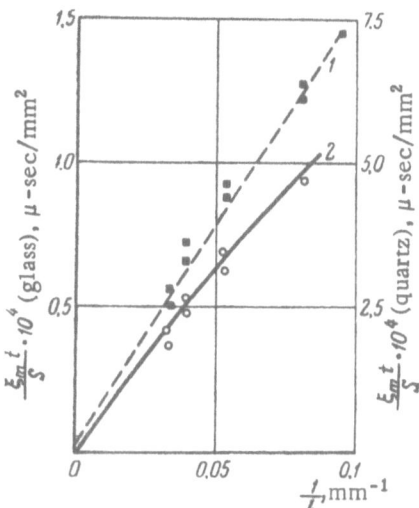

Fig. 53. Product of the amplitude and
the cutting time per unit area as a function of the reciprocal of the length of the
perimeter [44]: 1) glass; 2) quartz.

Veroman [36] has found that a grain loses its initial form within 10 sec; we have made similar measurements. A slurry of fresh boron carbide (120 mesh, Fig. 51a) was placed at the bottom of a blind hole 6 mm in
diameter and several mm deep, which had previously been made with the same tool. The slurry was removed
after a certain period of operation and was examined under the microscope. Figure 51b shows the grains and
the cut particles, from which it is clear that the volume cut is only a small fraction of the volume of the abrasive particles. Veroman found that the cut particles accounted for no more than 1% of the total, so any effect
of these on the cutting rate can be neglected.

The main factors are the concentration and particle size; the latter decreases substantially after a few
seconds, and the fall is pronounced even after the minimum practical time of 0.5–0.6 sec (Fig. 51c). Avery'yanova
and Milovidov [29] have concluded from high-speed cinematography that the effective lifetime is 0.5 sec.

The general mechanism of the process would lead us to expect that the rate of change of particle size is
governed by the number of particles in the gap, by the strength of the particles, by the amplitude, and by the
static force.

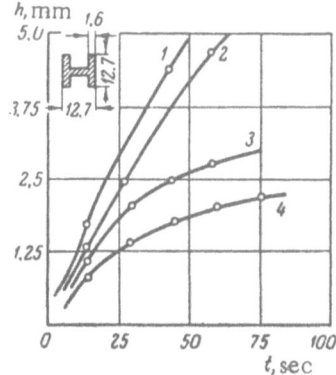

Fig. 54. Depth of hole as a function
of time for boron carbide of the following mesh numbers [8]: 1) 80;
2) 320; 3) M20; 4) M14.

The replacement of the abrasive in the gap has
hardly been examined, but it has been established that
increase in depth or area retards the influx. The conditions for breakage are not thereby altered, so the effective
grain size must become smaller; in fact, this is found to
be so (Fig. 52). Comparison with the cutting rate as a
function of depth shows that the rate is governed by the
grain size, which itself is controlled by the rate of influx.
The rate at any instant is determined by the balance between breakage and influx of new material, and it is constant unless the change in depth or area alters the rate of
influx.

Influx is therefore the rate-limiting factor for a
given static pressure and amplitude; it controls the effects
of depth, shape, and area. Any change of pressure or amplitude of course affects the rate; for instance, a constant

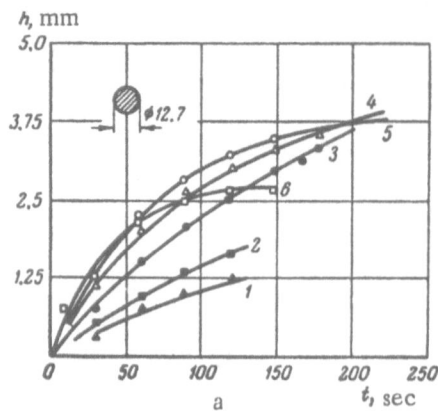

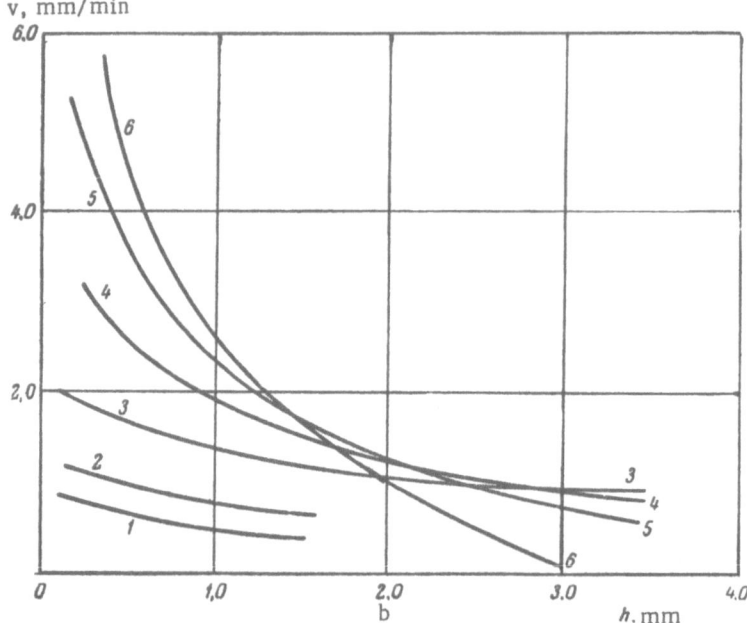

Fig. 55. Relation of a) depth to time and b) rate to depth [8] for F (kg)
as follows: 1) 0.23; 2) 0.45; 3) 0.9; 4) 1.6; 5) 2.7; 6) 3.6.

input power leads to a greater fall with depth than does a constant amplitude, because the amplitude falls in the first case. This is why results should be compared for a constant amplitude.

Barke [22] found that the cutting rate remained almost constant at 1.6 mm/min when glass was cut with tools of diameters 10, 15, and 20 mm (areas from 78 to 314 mm^2); the volume cut is therefore proportional to the area of the tool. These measurements were made at ξ_m = 10 μ and 1.5 kg/cm^2. However, the rate does not remain independent of the area for larger ranges; a small area gives a higher rate than a large one for the same amplitude [8].

The cutting rate is at first proportional to the area when a ceramic is cut with a hollow cylindrical tool, but a continuous fall soon sets in [12]. Further, the cutting rate is much higher for thin slots. It seems likely that in these cases there is some change in the cutting conditions apart from the effect of influx of abrasive.

Large changes in area cause a fall in volume rate as well as linear rate [7, 8]; we shall see that this is caused by effects on the flow of abrasive.

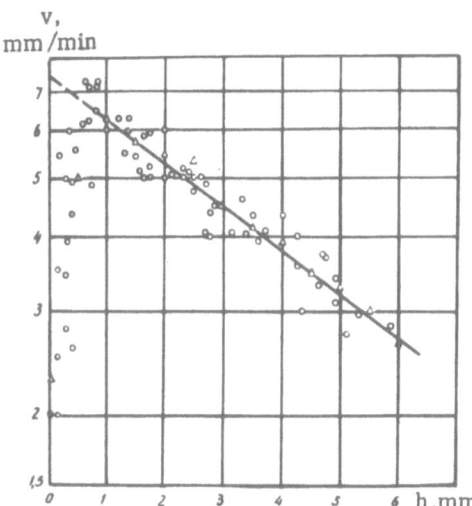

Fig. 56. Rate as a function of depth.

It has repeatedly been found that the time needed to form a hole of a given depth with a fixed amplitude increases with the area; moreover, the cutting rate falls more rapidly with depth the larger the area. Also, the mean rate is governed not only by the area but also by the shape of the hole; for instance, the rate for a long narrow rectangle is higher than that for a circle of the same area.

Goetze (see section 6) found for tool steel that the mean rate is proportional to the ratio of the area to the perimeter; Mizrokhi [44] found the same for glass and quartz (Fig. 53). Measurements were made of ξ_m and of t, the time needed to reach a depth of 4 mm. The product $\xi_m t/S$ (i.e., referred to the area of the tool) was compared with l, the length of the perimeter; Figure 53 shows that $\xi_m t/S \propto 1/l$. The depth remained constant at 4 mm, so the reciprocal of the time to 4 mm is proportional to the mean rate. The rate as a function of shape and size for small depths is not the same as that for large ones. The initial rate is governed mainly by the area of the tool, but the ratio of the area to the length begins to have a marked effect as the depth increases, for this controls the influx of abrasive.

The relation of depth to time for tools of various shapes has been given for cermets and hard alloys [45]. It was considered that the rate was independent of the area for small depths (the area ranged from 7 to 60 mm²); the rate fell as the depth increased, the fall being related to the size of the tool by $h = Ct^{0.8}$, in which h is depth, t is time, and C is a constant dependent on the shape and size of the hole.

This empirical relation involves the assumption that the curves are of the same shape under all conditions; but this is not so, for the grain size and static force also affect the shape.

Neppiras [8] found that the relation of depth to time is governed by the grain size; the smaller the latter, the more rapid the fall (Fig. 54).

This effect of grain size on the rate as a function of depth confirms that the rate of influx of abrasive is the decisive factor. The width of the gap between tool and side of hole is proportional to the grain size. Any increase in the length of this gap has a more pronounced effect on the influx for the smaller grain sizes, so the cutting rate falls more rapidly with depth for the smaller grain sizes.

It is usually [7, 8, 10, 22] assumed that the rate as a function of pressure has a peak governed by the amplitude, area, grain size, and so on; this is based on mean rates measured for depths of a few millimeters. Later, tests were done with optimal static forces, on the surprising assumption that this force is independent of depth, although the results indicate that the optimal pressure falls as the depth increases.

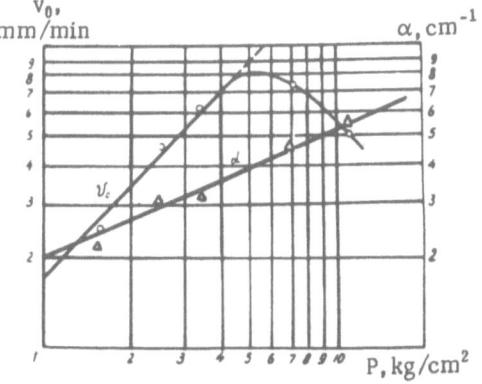

Fig. 57. Relation of α and v_0 in $v = v_0 \exp(-\alpha h)$ to pressure.

Figure 55a (from [8]) shows that the rate increases with the force, the highest rate corresponding to the highest force (3.6 kg); this applies to shallow holes 12.75 mm in diameter in glass. The variation in rate with depth is dependent on the force; different forces do not give curves of the same shape.

Figure 55a (curve 3) shows that the depth is almost proportional to the time at 0.9 kg, whereas 3.6 kg gives initially a higher rate; but then there is a falling off, the two curves meeting at a depth of about 2.5 mm. This means that the mean rate to that depth is the same; the larger force gives the higher mean rate for smaller depths.

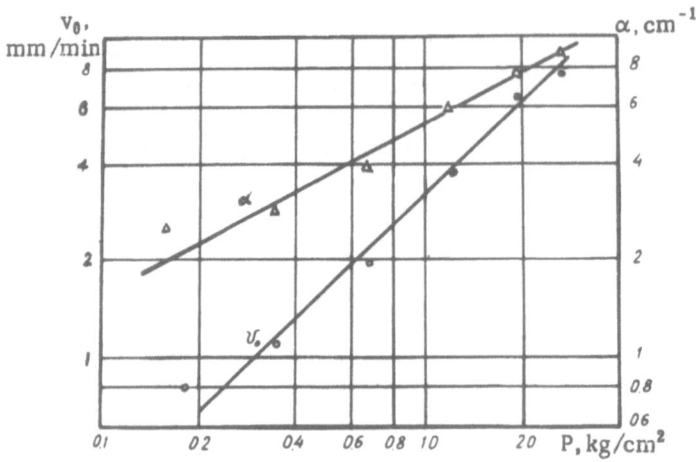

Fig. 58. Relation of Fig. 57 plotted from Neppiras's data
(Fig. 55).

Figure 55b shows the rate as a function of depth, which makes it clear that the rate falls off more rapidly with depth at the higher pressures.

We have performed similar tests with various tools on glass at various pressures with powders of various grain sizes to find the rate as a function of depth.

Figure 36 shows that the rate is initially somewhat low, a rise occurring around a depth of 1.0 mm and a continuous fall beyond this. A similar relation was dealt with in section 10 (Fig. 50a). All points for depths over 1 mm lie exactly on a straight line in logarithmic coordinates, which means that the rate at depth h is given by

$$v = v_0 e^{-\alpha h} \qquad \text{for } h > h_0 \tag{2.36}$$

in which v_0 and α are parameters.

The results show that this relation applies for any static pressure, though v_0 and α are functions of the pressure and the amplitude. Figure 57 shows these parameters as functions of the pressure from tests in which the pressure was the only variable.

These results show that the optimal pressure is a function of depth and so is not a constant. The best pressure must be found by experiment as that giving the minimum cutting time for a given depth. A substantial increase in over-all rate is available from a suitable program of pressure variation in some cases.

The results are fully explicable on the view that the balance between grain breakage and influx controls the cutting rate. The breakage rate is unimportant for small depths, on account of rapid replacement, so the force can be raised, which increases the cutting rate, because the force controls the stresses set up when the tool strikes the grains. We have seen in section 9 that the number and size of the chips are related to the mean force per particle.

Grain replacement becomes more difficult as the depth increases; so the breakage rate becomes more important; increased pressure raises the breakage rate, so the lack of abrasive becomes apparent at smaller depths. That is, the fall in rate with depth is more rapid for the higher forces. The shape and size of the hole also affect the variation in rate with depth, because a larger area in the lateral gap facilitates influx of fresh abrasive and so favors higher rates.

The cutting rate is therefore governed by the speed of migration in the lateral gap, so a study of this motion is of particular importance. There are two hypotheses on the mechanism whereby the particles reach the working gap. Firstly, it is supposed [18] that the slurry vibrates in accordance with the change in length of the tool. Secondly it is assumed [8] that the particles may be driven by stationary currents. The mode of motion may also be affected by microcurrents near the vibrating body [35, 36], though such motion has not been observed under actual conditions.

50

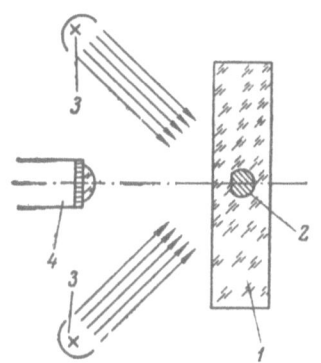

Fig. 59. System used to observe flow of abrasive slurry in the lateral gap.

Figure 59 shows the scheme we have used to examine the slurry in the lateral gap [46]. The glass block 1 was operated on by the tool 2, which has a flat front side. The particles could be observed through the side of the block. The light sources 3 and SKS-1 camera 4 were placed as shown, the speed being 1000 to 5500 frames per second at a magnification of two. The light sources were three SVDSh-250 mercury lamps arranged at 120° around a circle to illuminate the grains from all sides; the light was concentrated on the plane of the cut. The drive was provided by a 4770 machine (18 kc, amplitude up to 30 μ, 80 mesh abrasive).

The films showed that stationary flows play no part; the depth to which these extend is dependent on the motion of the slurry and on the method of feeding it to the cut. This depth in no case exceeded 0.5 mm, which is much less than the cutting depth, so the grains in the working gap take no part in this motion. The worn abrasive is replaced merely as a result of random motion of the particles. In most cases the stationary flows were not seen at all; there was merely random motion in the lateral gap, but this lack of stationary flows was almost without effect on the cutting rate. Only the random motion in the side clearance is responsible for replacement of abrasive. The abrasive under the end of the tool shows a similar motion (section 10); this is governed not by collision between the particles (which is rare), but by the motion of the liquid.

The speed of the particles increases with the amplitude, probably because the liquid in the side clearance moves at high speed when the tool vibrates. The particles are irregular in shape, so turbulence arises around them. The result is that in each half-cycle each particle is acted on by a force governed by the motion of the liquid and by the orientation of the particle.

Cavitation bubbles under the tool have a marked effect on the movement of the abrasive. These bubbles also move randomly, but much more rapidly than the particles; they tend to carry the latter with them, and these form a clump around each bubble. The vigorously pulsating bubbles also tend to scatter the particles and so facilitate mixing.

The mean speed of a particle is therefore much less than the instantaneous speed; the random motion means that the probability of a particle entering the working area decreases as the path to be traversed lengthens. In fact, the rate of exchange is inversely proportional to the depth of the cut; this also explains why the cutting rate ceases to vary with depth when a directional flow of liquid is set up.

Any improvement in the supply of slurry to the working gap accelerates the cutting; replacement of a cylindrical tool by a conical one [47] raises the cutting rate by a factor 1.5. It has been found [13] that withdrawal of the slurry (280 mesh carborundum) through a central hole raises the volume cutting rate. These tests were done at 22.6 kc with a power of 1 kW; Figure 60 shows the depth of the hole (in glass) as a function of time and size of tool. The steady flow of slurry raises the cutting rate considerably. Similar results have been reported [13] for ceramics.

It has been found for several materials [48] that withdrawal causes a marked rise in cutting rate (see Table 28). The slurry is withdrawn through a central hole in the tool into a sump at reduced pressure (pressure difference nearly 1 atm). Further tests [49] showed that the cutting rate remained constant down to depths of 10 mm or so.

The conclusion is that the cutting rate is governed solely by the maximal stress when the rate of supply of slurry is sufficient, this stress being governed by the pressure and the amplitude (see section 9). The volume cutting rate is then proportional to the area, to the pressure, and to the square of the amplitude, which means that the linear rate should be independent of area, as is actually found for low pressures (Fig. 61). However, the pressure affects the balance between rate of breakage and rate of replacement; we would expect that breakage would be less important for smaller areas, so the optimal pressure should be higher. The cutting rate is governed by the extent of grain fracture, so the cutting rate should fall when the optimal pressure is exceeded.

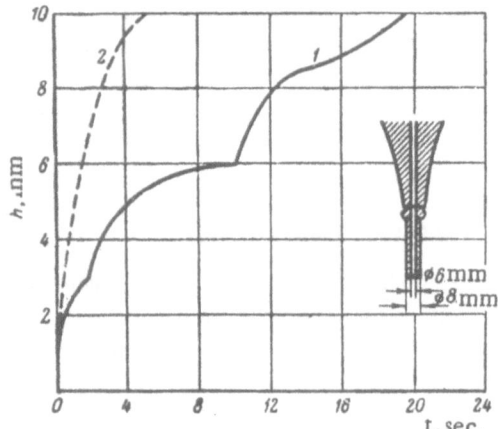

Fig. 60. Depth of hole as a function of time [13]: 1) without withdrawal of slurry; 2) with withdrawal.

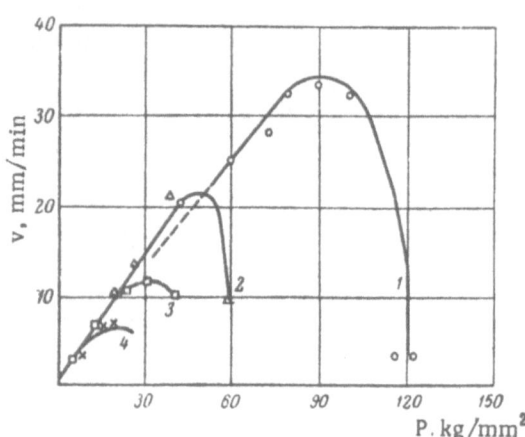

Fig. 61. Cutting rate as a function of pressure with slurry withdrawn through a central hole [49]. Outside diameter of tool D = 17 mm; inside diameter d (mm) as follows: 1) 15, 2) 13; 3) 9; 4) 3.

Grain breakage explains the observed [49] relation of linear cutting rate to amplitude (Fig. 62); the rate is proportional to the square of the amplitude for small amplitudes, but there is relatively little variation in rate for amplitudes over 30 μ.

This effect of slurry flow is very important, for it provides a great increase in cutting rate. The maximum possible rate should be attained if the flow rate is sufficient to replace the abrasive completely within one cycle.

The rate of flow of slurry to the working gap controls the cutting rate, so a detailed study should be made of the motion of the slurry with and without withdrawal.

The finish of the side surfaces is important, especially in precision machining; defects (channels or ridges) cause poor finish and accuracy. These defects on the tool and workpiece are caused by the movement of the abrasive in the gap.

Neppiras [8] supposed that fairly persistent flows may be set up over certain parts of the surface; particles carried by such flows mark the tool and hole, so channels are produced by prolonged cutting. Rapid cutting does not allow time for these flows to build up, so improved side finish occurs at high cutting rates. Fresh grooves are not usually produced by a second stage of machining, but existing ones are difficult to remove, for they act as ready-made channels for these flows.

Veroman [36] has described the main types of defect. The simplest is a longitudinal groove. He considers that cavitation produces grooves on the tool, whereas we believe that the slurry plays a major part in this. Any such groove is reproduced on the workpiece as a ridge, whose height may be 0.2 mm for hard alloys. Figure 63a shows a tool with grooves, and Fig. 63b the workpiece produced by such a tool. It has been shown* that these grooves are caused by bending vibrations in the tool, which cause the largest particles to move up and down with the tool and so form grooves. The liquid then flows in these grooves.

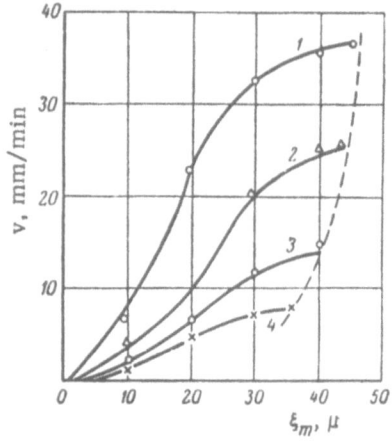

Fig. 62. Cutting rate as a function of amplitude with slurry withdrawal [49]. Outside diameter of tool D = 17 mm; inside diameter d (mm) as follows: 1) 15; 2) 13; 3) 9; 4) 3.

* These vibrations also make the hole elliptical, increase the side clearance, and reduce the cutting rate [49].

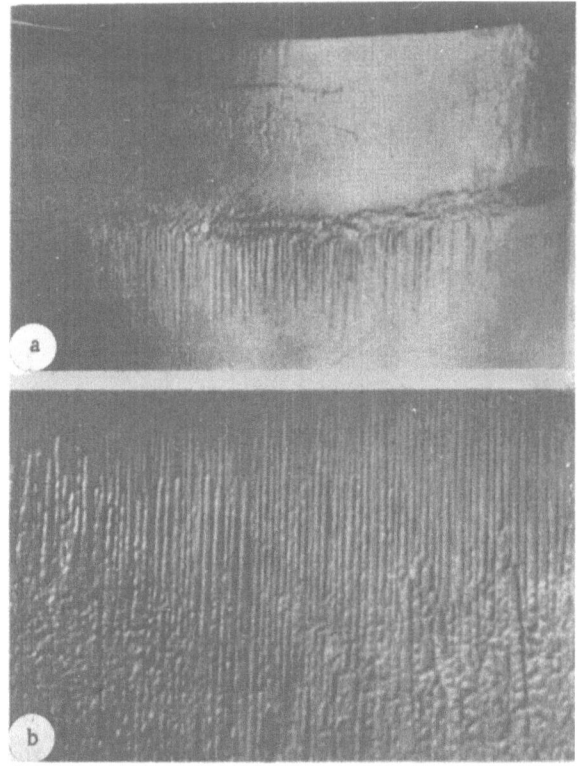

Fig. 63. Side surfaces of a) tool and b) workpiece.

Direct erosion of the workpiece also assists in producing grooves; the commonest form of this is a narrow horizontal groove 1 to 3 mm from the exit hole. This groove is seldom more than 0.1 mm deep or 0.3 mm wide. Such grooves sometimes occur all the way down.

Another type of defect has no very obvious direction, being randomly distributed over most of the surface but not found near the entrance hole. These pits are not so deep as the grooves. Figure 63b (lower part) shows these pits, whose form is such as to show that they are produced directly by cavitation.

Little is known about the causes and prevention of these defects, although they affect the surface finish. Future work should be directed to the relation of pitting to the motion of the slurry; lateral wear of the tool evidently has the same cause.

The following conclusions may be drawn from the evidence presented in this chapter.

The cutting rate is proportional to the static force and to the square of the amplitude, up to a certain limit; Neppiras's work on brittle materials has shown that the rate is proportional to the square root of the frequency. The effects of grain size on the rate and on the surface finish have been established; the rate and the surface roughness increase with the grain size, but there is a limit to the effect of the grain size on the rate, and a very coarse powder may even cause a fall in rate. The optimal grain size is governed by the amplitude. The cutting rate and maximal depth are dependent on the shape of the tool.

High-speed cinematography enables one to determine the basic mechanism of the process; the workpiece is cut only when the tool strikes directly on particles lying on the surface, which give rise to pits much smaller than the particles. The volume of material thereby removed is governed by the stress set up by the impact and by the range of shapes and sizes in the grains. The range of grain sizes means that the number of pits formed per blow is much less than the number of grains; this is in agreement with the observation that ultrasonic cutting is very wasteful of energy.

Cavitation plays no important part in the cutting, though cavitation erosion is responsible for much of the damage to the tool, whose surface finish is impaired, which is reflected in the surface finish of the workpiece. Cavitation reduces the amount of abrasive in the working gap, so the cutting rate is reduced.

Replacement of broken grains by new ones is of particular importance; the variation in cutting rate with depth is primarily governed by the rate of replacement, but the precise mechanism has not yet been established. It is supposed that random motion of the particles in the lateral gap is the principal cause of replacement; this motion also appears to control the damage to the side surface.

The practical uses of ultrasonic machining have determined the studies that have been made; the basic laws have been derived purely empirically, and there is no general theory of the process. A full study of the laws of erosion should undoubtedly lay the basis for further advances in ultrasonic machining.

Literature Cited

1. I. Emerson, "Ultraschall zur Metallbearbeitung," Am. Machinist 95(9):85, 1951.
2. S. G. Kelley, "Hard brittle materials machined using ultrasonic vibrations," Mater. Methods 34(3):92 1951.
3. G. H. De Groat, "Soft steel and ultrasonics machine carbide," Am. Machinist 15(9):141, 1952.
4. S. Spring,"Ultraschall-Bearbeitungsverfahren," Werkstatttechnik u. Maschinenbau 43(7):325, 1953.
5. M. S. Hartley, "Ultrasonic machining of brittle materials," Electronics 29(1):132, 1956.
6. N. J. Clark, "An ultrasonic machine tool," Trans. IRE, Nov. 10, 1954.
7. A. Nomoto, "Ultrasonic machining by low power vibration," J. Acoust. Soc. Am. 26(6):1081, 1954.
8. E. A. Neppiras, "Report on ultrasonic machining," Metalwork. Product. 100(27):1283,1956; 100(28):1333, 1956; 100(29):1377, 1956; 100(30):1420, 1956; 100(31):1464, 1956; 100(33):1554, 1956.
9. E. A. Neppiras and R. D. Fosket, "Ultraschall-Materialbearbeitung," Philips' tech. Rundschau 19(2):37, 1957.
10. P. E. D'yachenko, Yu. N. Mizrokhi, and V. G. Aver'yanova, "Some aspects of ultrasonic machining," in: Use of Ultrasonics in Industry, Moscow, Mashgiz, 1959, p. 149.
11. G. N. Kuzyaev and G. A. Tsveiman, "Ultrasonic apparatus for machining hard and brittle materials," Advanced Scientific and Industrial Experience, 9(365/4), 1957.
12. B. N. Lyamin, "Technical features of ultrasonic machining of metals," in: Use of Ultrasonics in Industry, Moscow, Mashgiz, 1959, pp. 136-145.
13. B. I. Mezhuev, "Ultrasonic machining of cermet materials," in: New Developments in Electrical and Ultrasonic Machining, Leningrad, 1959, p. 209.
14. N. Fukumoto, "Improvement of methods of ultrasonic mechanical machining," Machinery (Japan) 22(8): 1281, 1959.
15. D. Goetze, "Effect of vibration amplitude, frequency and composition of the abrasive slurry on the rate of ultrasonic machining in ketos tool steel," J. Acoust. Soc. Am. 28(6):1053, 1956.
16. D. Goetze and G. E. Miller, WADC Techn. Rept. 55-277, US Air Force, June, 1957.
17. D. Goetze, "Effect of pressure between tool, tip and workpiece on the rate of ultrasonic machining in ketos tool steel," J. Acoust. Soc. Am. 29(4):426, 1957.
18. G. E. Miller, "Special theory of ultrasonic machining," J. Appl. Phys. 28(2):149, 1957.
19. M. C. Shaw, "Das Schleifen mit Ultraschall," Microtechnic 10(6):265, 1956.
20. B. V. Mott, Hardness Testing by Microindentation, Moscow, Metallurgizdat, 1960.
21. V. I. Dikushin and V. N. Barke, "Ultrasonic erosion and its dependence on the vibrational characteristics of the tool," Stanki i Instr. No. 5:10, 1958.
22. V. N. Barke, "A study of the oscillatory system of the vibrator in an ultrasonic machine," Dissertation, Moscow Machine-Tool Institute, 1958.
23. N. S. Goryachev, Manufacture of Hard-Alloy Dies by Ultrasonic Machining, Leningrad, LDNTP, 1959.
24. L. D. Rozenberg and V. F. Kazantsev, "The physics of ultrasonic machining of hard materials," Doklady Akad. Nauk SSSR 124(1):79, 1959.
25. L. D. Rozenberg and V. F. Kazantsev, "A study of the mechanism of ultrasonic machining by high-speed cinematography," Stanki i Instr. No. 5:20, 1959.
26. I. V. Metelkin and V. V. Metelkin, "Physical principles of ultrasonic machining," Mashinostroitel' No. 10:9, 1958.

27. L. Balamuth, Method and Means for Removing Material from Solid Body, US Patent No. 2580716, dated Jan. 11, 1951.

28. I. S. Vainshtok, "Some aspects of the physical basis of ultrasonic machining," Stanki i Instr. No. 4:13, 1958.

29. V. G. Aver'yanova and A. A. Milovidov, "Study of machining of materials with ultrasound," in: Summaries of Papers at the Second Conference on High-Speed Photography and Cinematography," Moscow, Izd. Akad. Nauk SSSR, 1960.

30. V. P. Pukh, "A study of crack growth rates in transparent bodies by high-speed photography," in: Summaries of Papers at the Second Conference on High-Speed Photography and Cinematography, Moscow, Izd. Akad. Nauk SSSR, 1960.

31. V. F. Kazantsev, "Physics of ultrasonic machining of brittle materials," in: Use of Ultrasonics in Machine Construction, Moscow, 1960, p. 177.

32. G. Nishimura and S. Shimakawa, "Ultrasonic mechanical machining (part IV)," J. Fac. Eng., Univ. Tokyo 25(4):213, 1958.

33. G. Nishimura, K. Yanagishima, and T. Shima, "Ultrasonic mechanical machining (part VII)," J. Fac. Eng., Univ. Tokyo 26(1):53, 1959.

34. D. Blank, "Glasbearbeitung durch Stossläppen bei Ultraschallfrequenz (Ultraschallbearbeitung)," Glastech. Ber. 34(11):534, 1961.

35. V. F. Kazantsev, "Relation of cutting rate in ultrasonic machining to cutting conditions," Stanki i Instr. No. 3:12-15, 1962.

36. V. Yu. Veroman, "Ultrasonic method of making hard-alloy dies," in: Advanced Scientific and Industrial Experience 9(29/2), 1960.

37. F. J. Jackson, "Sonically induced microstreaming near a plane boundary. II. Acoustic streaming," J. Acoust. Soc. Am. 32(11):1387, 1959.

38. T. Norpo, Method of ultrasonic machining, Japanese patent No. 398, class 74NO, dated Jan. 24, 1957.

39. N. M. Rostovtsev, "Role of cavitation in the ultrasonic machining of hard bodies," Doklady Akad. Nauk SSSR 127(6):1210, 1959.

40. N. M. Rostovtsev, "Experience with ultrasonic machining of materials at high hydrostatic pressures," in: Use of Ultrasonics in the Examination of Materials, No. XII, Izd. MOPI, 1960, p. 53.

41. V. O. Mal'chonok and I. A. Utkin, "Effects of high external pressure on damage to the material in sonic machining of hard bodies," Akust. Zh. 6(1):128, 1960.

42. A. S. Bebchuk, "Cavitation damage to hard bodies," Akust. Zh. 3(1):90, 1957.

43. V. F. Kazantsev and Yu. L. Tissenbaum, "A study of the relation of rate of ultrasonic machining to temperature," Akust. Zh. 7(2):260, 1961.

44. Yu. N. Mizrokhi, "Ultrasonic machining of hard and brittle materials," in: Second Conference of Graduate Students and Junior Scientific Workers, Vol. 2, Trudy IMASh Akad. Nauk SSSR, Moscow, 1959, p. 87.

45. V. N. Berezub, A. E. Potapenko, and E. E. Chistyakov, "A study of the ultrasonic method of grinding cutting tools," Vestnik Mashinostr. No. 3:67, 1961.

46. V. F. Kazantsev and Yu. L. Tissenbaum, "Mode of motion of the abrasive in ultrasonic machining," Akust. Zh. 7(4):493, 1961.

47. V. V. Metelkin, "Ultrasonic machining of deep holes," Mashinostroitel' No. 8:28, 1960.

48. "L'usinage des métaux et corps durs sur la machine ultra-sonore Diatron," Ind. franc. Achats et entrét. matér. industr. 7:78, 891, 895, 897, 899, 900, 1958.

49. G. Pahlitzsch and D. Blanck, "Fortschritte beim Stossläppen mit Ultraschallfrequenz (Ultraschallbearbeitung)," Werkstattstechnik u. Machinenbau 50:592, 1960.

50. O. Nishimura, K. Yanagishima, and T. Shima, "Ultrasonic mechanical machining. IX. Machining speed and mixing ratio of abrasives," J. Fac. Eng., Univ. Tokyo, 26(2):129, 1959.

51. L. A. Shreiner, Hardness of Brittle Bodies, Izd. Akad. Nauk SSSR, Moscow, 1949.

52. M. I. Koifman, "Strength of hard mineral particles," Doklady Akad. Nauk SSSR 29(7):477, 1940.

53. M. Ohira, M. Kageyama, and O. Akutsu, "Study of ultrasonic machining. Contact angle, machining load, and penetration depth," J. Soc. Precis. Mech. Japan 27(7):480, 1961.

CHAPTER 3

THE ACOUSTIC SECTION
OF AN ULTRASONIC MACHINE TOOL

12. Relations between the Main Features of the Acoustic Head

A major working part of such a machine is the acoustic head (Fig. 64). The main function of this is to produce the vibration in the tool. The energy is drawn from the generator 1 in electrical form and is converted to mechanical form by the transducer (vibrator or radiator) 2, which periodically shortens and lengthens. The amplitude of this motion is usually inadequate for cutting purposes, so the vibrator is joined to a concentrator 3, which is simply a convergent waveguide designed to produce the desired amplitude at the far end. The two together form the vibrating system, which is loaded at the far end. The resistive component of the load represents the energy lost to the gap 6 between the tool 5 and workpiece 4. The mass of the tool 5 forms the reactive part.

The main channel for the ultrasonic energy is formed by the vibrator, concentrator, and load; the main requirements the channel must satisfy are that the ultrasonic vibrations must be of sufficient power and that the energy must pass freely from vibrator to load with minimal loss on the way.

The vibrator is made resonant in order to obtain a sufficient amplitude; the length in the direction of propagation is made a half wavelength (or occasionally a multiple of this). The head is supplied with a frequency such as to ensure this.

The concentrator is also made resonant; it becomes a volume resonator tuned to the same frequency, which provides the best conditions for power transfer.

There is always a second channel for the energy, which is formed by the holder 7 and fixed parts 8 (by fixed we mean, here and later, parts that do not participate in the working vibration; mechanically speaking, they are not fixed, because they are usually linked to the feed mechanism). Here the requirements are entirely different; this part must receive the minimum amplitude, and the impedance must be reactive, in order to avoid energy loss.

The two channels are linked; the first consists of the resonant elements and the load, in which the traveling wave is dominant, if the transfer to the load is efficient. The second must be of minimal loss, which means that the fixed components must lie at the nodes of the standing waves in the first channel. We shall see in section 16 that the attachment may be made directly at the nodes or via special holding components, whose sizes and positions are strictly related to the standing-wave pattern.

Now we consider the effects of the load. The resistive load must be matched to the vibrating system as closely as possible in order to obtain the maximal traveling-wave coefficient. The reactive component (especially that arising from the mass of the tool) must be minimized, for this component affects the phase distribution in the standing waves and may tend to increase the loss to the fixed parts.

Nonlinear features can also adversely affect the operation; for instance, nonlinearity in the vibrator may result in the production of a spectrum, which implies energy loss, for the resonant elements and mode of attachment are designed to operate only over a narrow range of frequencies, so the additional frequencies are not utilized efficiently. Nonlinearity in other parts is also disadvantageous.

Absorption in any element also reduces the efficiency; we have seen in Chapter 1 that the main cause of absorption in most solids is internal friction. The loss is approximately proportional to the square of the deformation rate and to the volume. Consider a rod whose cross section S is a function of x and examine the effects

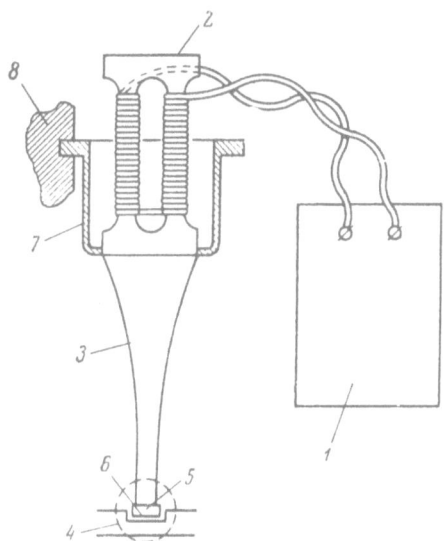

Fig. 64. Simplified diagram of the acoustic head.

in an element of thickness dx (Fig. 74); a longitudinal sinusoidal elastic wave causes the absorption of the energy

$$dW = \frac{\omega^2}{E^2} \sigma^2(x, t)\,\theta\; S(x)\,dx, \tag{3.1}$$

in which $\omega = 2\pi f$ is angular frequency, E is Young's modulus, σ is stress, and θ is the loss coefficient.

For the length l we have the total loss* as

$$W = \frac{\omega^2\theta}{2E^2} \int_{C}^{l} \sigma^2(x)\,S(x)\,dx. \tag{3.2}$$

Sometimes it is useful to characterize the loss by the mechanical loss resistance, which is defined by

$$W = \frac{1}{2}\, v_{m1}^2\, R \tag{3.3}$$

in which v_{m1} is the amplitude of the velocity at the point for which the loss resistance is to be defined. The mechanical loss resistance at the point is

$$R = \frac{\omega^2\theta}{E^2 v_{m1}^2} \int_{0}^{l} \sigma^2(x)\,S(x)\,dx. \tag{3.4}$$

It is convenient to relate this to Q, which is readily measured. We have

$$Q = \frac{\omega M_e}{R}, \tag{3.5}$$

in which M_e is some equivalent mass for the section for which R is to be found. To find M_e we take the kinetic energy of an element of volume

$$dK = \frac{1}{4}\,\rho v^2(x)\,S(x)\,dx, \tag{3.6}$$

---

* Here and later the time-averaging of harmonically varying parameters (velocity, stress, and so on) is assumed.

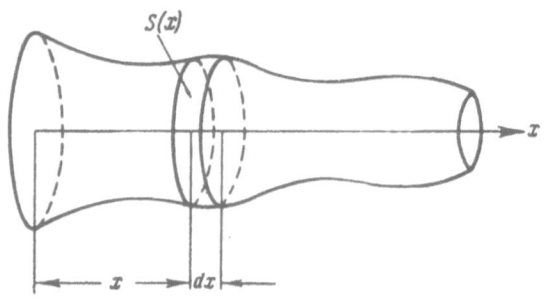

and find the total energy as

$$K = \frac{1}{4}\,\rho \int\limits_0^l v^2\,(x)\,S\,(x)\,dx. \tag{3.7}$$

On the other hand, for this section we have

$$K = \frac{1}{4}\,v_{m1}^2\,M_e. \tag{3.8}$$

which gives

$$M_e = \frac{\rho}{v_{m1}^2} \int\limits_0^l v^2\,(x)\,S\,(x)\,dx. \tag{3.9}$$

Fig. 65. Derivation of the basic equations
for a nonuniform rod.

Then with (3.5),

$$Q = \frac{\rho E^2}{\omega \theta}\;\frac{\int\limits_0^l v^2\,(x)\,S\,(x)\,dx}{\int\limits_0^l \sigma^2\,(x)\,S\,(x)\,dx}\;. \tag{3.10}$$

We substitute into (3.4) the θ of (3.10) to get the loss resistance as

$$R = \frac{\omega \rho}{Q v_{m1}^2} \int\limits_0^l v^2\,(x)\,S\,(x)\,dx. \tag{3.11}$$

It is convenient here to introduce the velocity distribution normalized with respect to x by reference to the fixed velocity v_{m1}:

$$v_N\,(x) = \frac{v\,(x)}{v_{m1}}\,. \tag{3.12}$$

In normalized form we have

$$R = \frac{\omega \rho}{Q} \int\limits_0^l v_N^2\,(x)\,S\,(x)\,dx. \tag{3.11a}$$

It is simple to measure ρ, ω, and Q; the integrand is dependent on the geometry of the rod and on the mode of normalization.

This discussion of the linear mechanical loss is essential to the consideration of the various rod components of acoustic nodes.

The close coupling of the components necessitates this discussion of energy transfer. Some features of the behavior of the various components are very similar and can be described by a single mathematical treatment.

13. Electromechanical Transducers

Magnetostriction transducers (see [1-4]) are used in nearly all current ultrasonic machine tools; their undoubted advantages are their high efficiency and high reliabililily in the range 15-30 kc, as well as the low supply voltages they need, which simplify cooling. The magnetostriction effect is one in which the material changes in dimensions in response to a magnetic field; it occurs to some extent in all ferromagnetics. Nickel, for example, contracts (negative magnetostriction), while an iron-aluminum alloy (alfer) expands (positive magnetostriction).

A specimen of magnetostrictive material in an alternating field vibrates at twice the frequency of the field, as follows from the symmetry of the magnetostriction curve (Fig. 66). The slope near zero is small, so the amplitude is very small; the system is then nonlinear, which increases the loss. All these undesirable effects

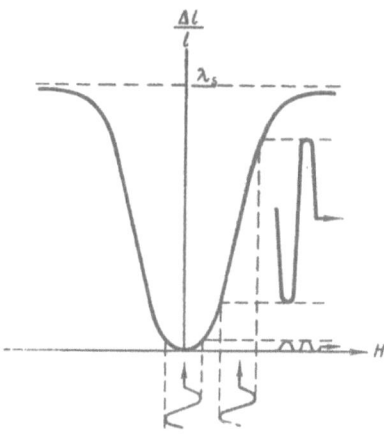

Fig. 66. Characteristic curve
of magnetostriction.

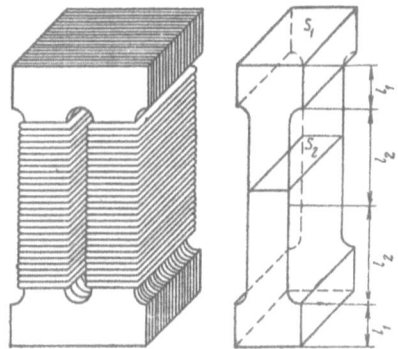

Fig. 67. A magnetostriction transducer.

are suppressed by bias magnetization; a steady field is superimposed on the alternating one, which displaces the working point to a linear part of the curve.

Neglecting effects of hysteresis type, we consider a rod of cross section S working on the linear part of the magnetostriction curve in response to an alternating field whose induction is B and which produces an elastic force

$$F_M = aSB,\qquad(3.13)$$

in which a is a constant. The converter experiences a stress

$$\sigma_M = aB.\qquad(3.14)$$

These relations are, in general, inexact. A resonant rod has intense standing waves, and large elastic deformations always occur at the middle of a half-wave rod, so in this area we get the inverse effect (weakening of the field by the stress). The above expressions then apply only far from such points.

A practical transducer is usually made of thin plates, in order to minimize losses; Fig. 67 shows a typical form, in which the electrical energy is converted to mechanical form via coils, which usually carry the ac as well as the dc (bias current). The horizontal ends serve to close the magnetic circuit and also act as the emitting surface.

For future purposes it is convenient to assume that the force given by (3.13) is applied at the point where the vertical and horizontal parts join.

Exact calculations [5-7] are rather complicated; in fact, no method suitable for engineering use exists, and it is usual to use more or less gross approximations. On the other hand, there is a feature that simplifies matters in the design of machines, in that the radiator can be made in accordance with approximate methods and tested for resonance frequency, after which the rest of the acoustic part can be designed to suit.

It is often sufficient to calculate only for the side parts as loaded by the ends (Fig. 67), in order to design the radiator roughly to a given frequency. The usual symmetrical system allows one to use the wave equation for uniform rods (Chapter 1), which gives [8] the resonance condition as

$$\cot kl_1 \cot kl_2 = \frac{S_1}{S_2},\qquad(3.15)$$

in which S_1 and S_2 are the cross sections of a rod and of the part of the end piece assigned to the rod; l_1 is the height of the end piece, l_2 is half the length of the rod, $k = 2\pi f/c$ is the wave number, f is frequency, and c is the speed of sound in the material.

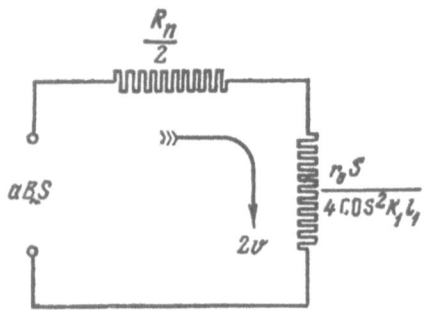

Fig. 68. Equivalent mechanical circuit of a magnetostriction transducer in unilateral radiation at its resonant frequency.

From (3.15) we have for a given f and c

$$2l_2 = \frac{c}{\pi f}\ \arctan\left(\frac{S_1}{S_2}\tan\frac{2\pi f l_1}{c}\right).\tag{3.16}$$

S_2, S_1, and l_1 are determined in accordance with energy and acoustic considerations. The material is specified, together with the radiated power; the maximum input power for given cooling conditions is known, so the area of the rod is determined. On the basis of optimal cooling, uniform magnetization, and convenient location of the coils, the number of rods is determined, which uniquely defines S_2. It is usually convenient to have the emitting surface square, so S_1 for the part of the end piece per rod is defined by the width of the winding area, i.e., by the coils and cooling needed. The height l_1 is deduced as follows. The end piece must not present a high magnetic reluctance; for instance, its height in a symmetrical system must not be less than the width of a rod, otherwise magnetic saturation may occur, which would cause a marked increase in the loss. Moreover, high induction in the end piece would produce magnetostriction in it, and the resulting oscillations would be perpendicular to those in the rod, which would increase the machining error as well as the losses. On the other hand, an unduly large height is also undesirable; a thick end piece reduces the volume of material in which useful conversion occurs.

Before we can make detailed electrical calculations, we must consider some important features of the operation under load.

We have seen in Chapter 1 that the behavior of a mechanical system operating at or near resonance is governed by the loss and load resistances. Figure 68 shows the approximate equivalent mechanical scheme [6] containing a magnetostriction force together with resistances (loss and load). The velocity (Chapter 1) is found by dividing the force by the sum of the resistances. Let v be the velocity at the emitting surface, $B_\sim$ the alternating component of the induction in the rods, S_t the total cross section of all rods, a the magnetostriction constant, $k_1 = 2\pi f/c$ the wave number for the material of the radiator, l_1 the height of the end pieces, r_0 the specific resistance of the load, S the area of the radiating surface, and R the mechanical-loss resistance for the convertor (section 12). Now a = constant for a given material, if the devices work over the linear part of the magnetostriction curve. If, instead, the linear part of the magnetization curve is used, we must make the substitution

$$aB_\sim = (\varkappa\Gamma)H_\sim,\tag{3.17}$$

in which $H_\sim$ is the strength of the alternating field and $(\varkappa\Gamma)$ is a certain magnetostriction constant [1], in which a and $(\varkappa\Gamma)$ are related by $a = \varkappa\Gamma/\mu$, in which μ is the magnetic permeability. Of course, hysteresis is neglected.

The equivalent circuit gives *

$$v = \frac{2aB_\sim S_t\ \cos^2 k_1l_1}{r_0 S}\ \frac{1}{1+\dfrac{2R\ \cos^2 k_1l_1}{r_0 S}},\tag{3.18}$$

so the amplitude is given by (3.3) as

$$v_m = \frac{2(\varkappa\Gamma)H_m S_t\ \cos^2 k_1l_1}{r_0 S}\ \frac{1}{1+\dfrac{2R\ \cos^2 k_1l_1}{r_0 S}}.\tag{3.19}$$

Chapter 1 gives the total radiated power as

$$W_a = \frac{1}{2}v_m^2 Sr_0 = \frac{2(\varkappa\Gamma)^2 H_m^2 S_t^2\ \cos^4 k_1l_1}{r_0 S}\ \frac{1}{\left(1+\dfrac{2R\ \cos^2 k_1l_1}{r_0 S}\right)^2}.\tag{3.20}$$

_____

* An error occurs in the corresponding calculations in [6]; the expression for the acoustic mechanical efficiency [the second factor in our (3.18)] does not have the specific load resistance multiplied by the radiating area.

On the other hand, Chapter 1 shows that $v = \omega\xi$ for harmonic vibrations, and the same acoustic power may be expressed via the amplitude of the displacement at the surface:

$$W_a = \frac{4\pi^2 f^2}{2}\xi_m r_0 S. \tag{3.21}$$

We equate the right-hand sides of (3.20) and (3.21) to get a formula for the total area of the rods:

$$S_t = \left(1 + \frac{2R}{r_0 S}\frac{\cos^2 k_1 l_1}{}\right)\frac{\pi f \xi_m r_0 S}{(\varkappa\Gamma) H_m \cos^2 k_1 l_1}. \tag{3.22}$$

Before examining this we replace the quantities representing the acoustic field at the radiating surface by the corresponding quantities for the cutting area. Then

$$\left.\begin{array}{l} \xi_m / \xi_{mc} = \dfrac{1}{k}, \\[2mm] r_0 S = k^2 r_{0c} S_0 \end{array}\right\}, \tag{3.23}$$

in which ξ_{mc} is the displacement amplitude in the cutting zone, r_{0c} is the specific acoustic resistance of the load in that zone, S_0 is the area at the exit end of the concentrator, and k is the gain (see section 14). Then

$$S_t = \left(1 + \frac{2R}{k^2 r_{0c} S_0}\frac{\cos^2 k_1 l_1}{}\right)\frac{\pi f \xi_{mc} k\, r_{0c} S_0}{(\varkappa\Gamma) H_m \cos^2 k_1 l_1}. \tag{3.22a}$$

This expression shows that the area of the rods must increase with the frequency and amplitude; it must also increase as the induction that can be produced decreases. The larger $(\varkappa\Gamma)$ or a, the smaller and ligher the head, other things being equal.

The load parameters have rather complicated effects. If we assume that the converter itself has little or no mechanical loss (R = 0), we can put (3.22) in the form

$$S_t = \frac{\pi f \xi_{mc} k r_{0c} S_0}{(\varkappa\Gamma) H_m \cos^2 k_1 l_1}. \tag{3.24}$$

The area of the rods is then directly proportional to the specific resistance of the load, to the machining area, and to the gain of the concentrator; moreover, it is appreciably dependent on the dimensions of the end pieces, for the thicker these are the larger the area must be.

On the other hand, this approach is generally far from correct, for we know (see Chapter 2) that the actual cutting consumes only a very small fraction of the power; the load has little effect on Q, and nonlinearity in the load has quite inappreciable effects. This indicates that (3.24) is not applicable to actual systems; it is better to use (3.22a).

The importance of the losses can be established from another extreme case.

Let us suppose that the internal loss is so large relative to the radiation loss that the latter can be neglected in the first factor on the right in (3.22a); then

$$S_t = \frac{2\pi f \xi_{mc} R}{k (\varkappa\Gamma) H_m}. \tag{3.25}$$

Then S_t is independent of the load parameters; R is roughly proportional to the volume of material (i.e., to S_t), so this would mean that the cross section can be selected without reference to the other parameters of the process. This is not so, of course, for the true position lies somewhere between these extreme cases; the real case is described by (3.22a), (3.18), (3.19), and (3.20) in general form.

It is not very difficult to calculate R if Q is measured. For instance, R' (the R for one rod) is given by (3.11a), for which the integral of the square of the normalized velocity function may (Fig. 67) be conveniently split into four parts (symmetrical radiator), each the same definite integral:

$$I_N = \int_0^l v_N^2(x) S(x)\, dx = 2\int_0^{l_1} v_N^2(x) S_1 dx + 2\int_{l_1}^{l_1+l_2} v_N^2(x) S_2 dx. \tag{3.26}$$

I_N is found in accordance with the energy distribution, and the mechanical losses are referred to $x = l$ (junction of rod and end piece):

$$I_N = \frac{l_1 S_1 + l_2 S_2}{\cos^2 k_1 l_1} + (S_1 - S_2)\frac{\tan k_1 l_1}{k_1}, \tag{3.27}$$

whence

$$R = \frac{2\pi f\rho}{Q}\left[\frac{l_1 S_1 + l_2 S_2}{\cos^2 k_1 l_1} + (S_1 - S_2)\frac{\tan k_1 l_1}{k_1}\right]. \tag{3.28}$$

The total mechanical loss resistance R is found by multiplying R' by the number of rods.

It is usual to employ less sound but far simpler methods, or even to classify radiators in accordance with the acoustic power they can emit unilaterally into water; the wave impedance of this medium is

$$(\rho c)_W \cong 1.5 \cdot 10^5 \text{ g/cm}^2 \cdot \text{sec}.$$

This parameter is commonly employed in design; some simplification of (3.18) or (3.19) gives a formula for the total cross section of the rods:

$$S_t = W_a \frac{d(\rho c)_W}{2\eta^2 (\varkappa\Gamma)^2 H_m^2 \cos^4 k_1 l_1}, \tag{3.29}$$

in which W_a is the acoustic power transmitted to water and $d = S/S_t$ is a design constant governed by the width of the winding area (usually 1.2-1.5, width increasing with d); η is the efficiency, with allowance for the mechanical loss (usually 0.8-0.9; high loss, low η).

It is sometimes more convenient to use the amplitude of the magnetic induction, in which case we have

$$S_t = W_a \frac{d(\rho c)_W}{2\eta^2 a^2 B_m^2 \cos^4 k_1 l_1}. \tag{3.29a}$$

We have seen that a and $(\varkappa\Gamma)$ are related via μ for the actual working conditions; for instance, for nickel under the conditions usually used (H_0 of 10-15 Oe) we have

$$a = 9.4 \cdot 10^3 \text{ dyne/cm}^2 \cdot \text{gauss};$$
$$(\varkappa\Gamma) = 40 \cdot 10^4 \text{ dyne/cm}^2 \cdot \text{oersted}.$$

The units to be used with (3.29) or (3.29a) are cgs electrostatic, for which W_a is put as erg/sec (1 W = 10^7 erg/sec).

Such calculations are only approximate, but this serves to justify some of the assumptions, such as linearization of the magnetic or magnetostriction response within the working range, use of static characteristics instead of dynamic ones, and so on.

Figure 69 shows the static magnetization and magnetostriction characteristics for the most commonly used materials. The upper limit to $\Delta l/l$ corresponds to λ_s, the saturation magnetization (Table 5).

The following main requirements should be satisfied as far as possible when selecting the working point (see also [2]):

1) The magnetostriction curve should be linear and of maximal slope within the working range;

2) the working range should be as wide as is compatible with the fall at high and low values;

3) the effective magnetic permeability should be high, for the onset of saturation causes increased loss from magnetic leakage;

4) the linear part of the magnetization curve should be used, in order to minimize distortion and magnetic losses.

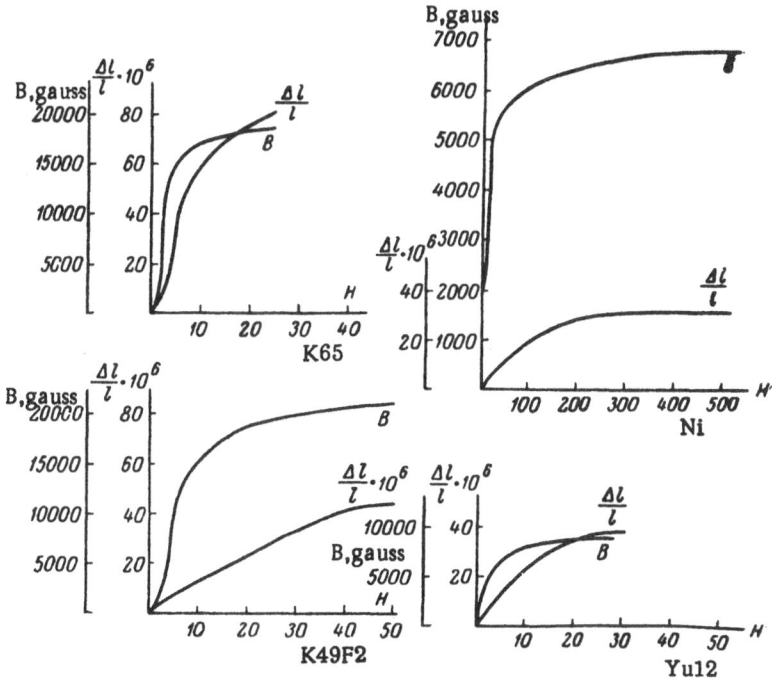

Fig. 69. Magnetostriction and magnetization curves
for the commoner magnetostriction materials.

The curves of Fig. 69 show that it is not possible to satisfy the third requirement for nickel for large amplitudes, so nickel units usually operate near saturation ($\mu \approx 42$). In general, nickel is not the best of materials, particularly in view of its rather low elastic limit, which gives rise to mechanical losses for amplitudes in excess of 3-5 μ. The main advantages of nickel are cheapness and simplicity of machining.

See [8] for further details of the purely electrical calculation. From the geometry of the converter and the effective voltage U provided by the source we have the number of turns as

$$n = \frac{U}{4.44 f S_{\text{м}} B_m} \cdot 10^8, \tag{3.30}$$

in which f is in cps, U is in volts, B_m is in gauss, and S_M is the cross section of the magnetic material in cm².

The specific losses (W/kg) from eddy currents and hysteresis are

$$\omega_e = \frac{\delta_1}{\rho} (df B_m)^2 \cdot 10^{-10} \tag{3.31}$$

and

$$\omega_h = \frac{\delta_2}{\rho} f \left(\frac{B_m}{1000} \right)^{1.6} \cdot 10^{-2}, \tag{3.32}$$

in which d is the thickness of the strips and δ_1 and δ_2 are coefficients, whose values are 1.16 and 0.75, respectively, for nickel. Then the total magnetic loss is

$$W_{\text{м}} = (\omega_e + \omega_h) G_{\text{м}}, \tag{3.33}$$

in which G_M is the weight of the magnetic material (nearly the total weight).

The effective current in the winding is given by

$$I = \sqrt{\left(\frac{W_a}{\eta U} + \frac{W_{\text{м}}}{U} \right)^2 + 0.57 \left(\frac{B_m l_{\text{м}}}{\mu\, n} \right)^2}, \tag{3.34}$$

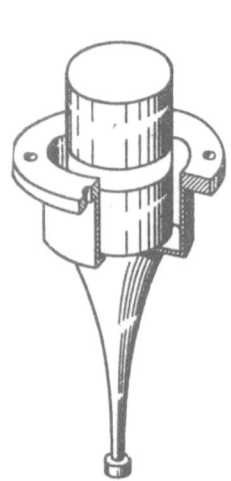

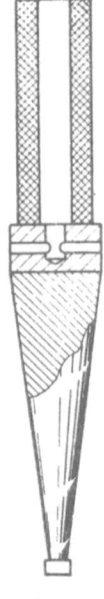

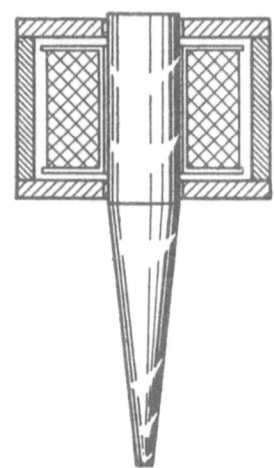

Fig. 70. Acoustic head with Langevain sandwich as its transducer.

Fig. 71. Acoustic head with cylindrical ceramic as its transducer.

Fig. 72. Acoustic head with transducer of magnetostrictive ferrite.

in which l_M is the length of the magnetic circuit. This then gives us the diameter of the wire to be used.

The final step is usually to calculate the efficiency of the device, but this is here a purely formal step, for the over-all efficiency is minute, and some parameters (impedance of load, mechanical losses, etc.) are not known precisely.

Magnetostriction is not the only means of producing ultrasonics for machining; Figure 70 shows the use of a Langevain sandwich, in which a piezoelectric plate is enclosed between two steel ones [9, 10].

The inverse piezoelectric effect is simply the capacity of a body to change its dimensions in response to an electric field; it has long been used as a means of generating ultrasonics [10]. Natural piezoelectrics are quartz and tourmaline; the best material for the purpose is a synthetic ceramic in which many small particles are bound together by sintering. Such a sinter must first be polarized by heating it above the Curie point and placing it in a strong electric field, which causes all the spontaneous-polarization vectors to point in the same direction, the orientation being preserved on cooling.

However, the main region of application of piezoelectrics lies above 300 kc. The velocity of sound in such a sinter is of the same order as that in a metal, so resonant plates having frequencies of 20-25 kc (the range of interest here) would be very thick, which would demand very high voltages; the materials are of low thermal conductivity and have fairly high mechanical and electrical losses, so they may readily become heated above their Curie points (about 125°C for barium titanate) at high amplitudes. In fact, they cannot provide large amplitudes, which is why Langevain proposed the use of a thin plate enclosed in metal plates, which serve to reduce the resonance frequency. Another possibility is to use cylinders of barium titanate with lengthwise vibrations [11], as shown in Fig. 71.

Piezoelectric devices have the advantages of simplicity and high rigidity, and they appear to hold much promise, especially since new ceramics of high efficiency based on lead titanate-zirconate and lead niobate have become available. Magnetostriction ferrites are also promising; these are synthetic polycrystalline materials of very low magnetic loss and hence of higher efficiency. Moreover, they enable one to make systems of almost any shape. The first ultrasonic machines employing ferrites (Fig. 72) have already been described [12].

14. Rod-Type Ultrasonic Concentrators

The main purpose of a concentrator is to increase the amplitude to the level needed for cutting; the concentrator is made as a tapered rod with a specified law of taper. Mason and Wick [13, 14] and Rozenberg and Lozinskii [15] described this independently.

One of the commonest types has an exponential law of variation:

$$S = S_0 e^{-\alpha x}.$$ (3.35)

The propagation of a longitudinal wave in such a rod is described by Webster's equation* :

$$\frac{\partial^2 \varphi}{\partial x^2} + \alpha \frac{\partial \varphi}{\partial x} - \frac{1}{c^2} \frac{\partial^2 \varphi}{\partial t^2} = 0,$$ (3.36)

in which φ is the velocity potential, x is coordinate along the rod, and c is the velocity of sound in the material. The solution for harmonic waves in an exponential rod is

$$\varphi = \varphi_m e^{\frac{\alpha}{2}x}(e^{-j\frac{\omega x}{c'}} + \beta e^{j\frac{\omega x}{c'}})\, e^{j\omega t};$$ (3.37)

in which $\omega = 2\pi f$ and $c' = \dfrac{c}{\sqrt{1 - \dfrac{\alpha^2 c^2}{4\omega^2}}}$ is the phase velocity of sound in the rod; β is the reflection coefficient at the end. The mechanical stress and velocity are then

$$\sigma = \rho \frac{\partial \varphi}{\partial t} = j\omega\rho\varphi_m e^{\frac{\alpha}{2}x}(e^{-j\frac{\omega x}{c'}} + \beta e^{j\frac{\omega x}{c'}})\, e^{j\omega t},$$ (3.38)

and

$$v = -\frac{\partial \varphi}{\partial x} = -\varphi_m e^{\frac{\alpha}{2}x}\left[\left(\frac{\alpha}{2} - j\frac{\omega}{c'}\right)e^{-j\frac{\omega x}{c'}} + \beta\left(\frac{\alpha}{2} + j\frac{\omega}{c'}\right)e^{j\frac{\omega x}{c'}}\right]e^{j\omega t}.$$ (3.39)

The rod is usually resonant, so its length is

$$L = n\frac{\lambda_c}{2},$$

in which $\lambda_c = c'/f$ is the wavelength in the concentrator; then at x = 0 we have

$$\sigma_0 = j\omega\rho\varphi_m\,(1+\beta)e^{j\omega t},$$

$$v_0 = -\varphi_m\left[\left(\frac{\alpha}{2} - j\frac{\omega}{c'}\right) + \beta\left(\frac{\alpha}{2} + j\frac{\omega}{c'}\right)\right]e^{j\omega t};$$

and for x = L = nc/2f

$$\sigma_L = j\omega\rho\varphi_m e^{\frac{\alpha}{2}L}(e^{-jn\pi} + \beta e^{jn\pi})\, e^{j\omega t},$$

$$v_L = -\varphi_m e^{\frac{\alpha}{2}L}\left[\left(\frac{\alpha}{2} - j\frac{\omega}{c'}\right)e^{-jn\pi} + \beta\left(\frac{\alpha}{2} + j\frac{\omega}{c'}\right)e^{j\omega \pi}\right]e^{j\omega t}.$$

For any integral n

$$\frac{v_L}{v_0} = \frac{\sigma_L}{\sigma_0} = \mp e^{\frac{\alpha}{2}L} = \pm\sqrt{\frac{S_l}{S_0}},$$ (3.40)

in which S_l is the cross section at the exit end.

---

* Webster averaged the acoustic potential over the cross section in deriving this, so it is not applicable when the transverse dimensions are greater than half a wavelength [16]. Moreover, no allowance was made for mechanical losses.

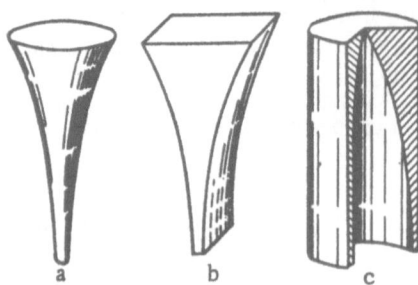

Fig. 73. Exponential concentrators:
a) circular; b) wedge; c) annular.

The method of designing the rod is clear from these expressions. Similar calculations are possible if the rod is not resonant. This is sometimes necessary if the concentrator, although resonant as a whole, is to be built up from several parts with different values of α. The following is the method for an exponential part in most general form.

We specify f, S_0, and S_l, as well as c and m (the length of the part in terms of wavelengths). The area ratio is

$$N = \sqrt{\frac{S_0}{S_l}},$$ (3.41)

and equals the gain in displacement, stress, or velocity for an exponential rod. The geometrical length is

$$L = \frac{mc}{f} \sqrt{1 + \left(\frac{\ln N}{2\pi m}\right)^2},$$ (3.42)

and the exponent is

$$\alpha = \frac{2 \ln N}{L},$$ (3.43)

so the variation in cross section is

$$S = S_0 e^{-\alpha x}.$$ (3.44)

This variation in cross section can be brought about in various ways; the rod or its parts may take any of the forms shown in Fig. 73, all of which give the required concentration for different purposes.

An ordinary tapering circular rod has an area proportional to the square of the diameter; an exponential wedge has one side of constant length, while a hollow rod has an annulus of variable width given by

$$S = \frac{\pi}{4} (D^2 - d^2),$$ (3.45)

in which D is the outside diameter and d is the inside diameter. This formula, or the corresponding one, must be inserted in (3.41)-(3.44) for detailed calculations.

Apart from exponential forms, studies have been made on straight conical, catenary, and stepped forms (Fig. 74). There are many papers [17-24] on the design of these; we merely quote the basic formulas (Table 13), which can be adapted to particular cases. Exponential and symmetrical stepped forms are the commonest of those listed in Table 13; each type has its advantages (simplicity and high gain in the stepped type, better matching in the exponential), and the preference for these forms arises from the ease of sufficiently precise calculations for them.

The rounding-off of the sharp corner between sections in the stepped type has very little effect on the result; such rounding is necessary to minimize stress concentration, which can lead to overheating and fracture. Such concentrators do not give the calculated gain if the area ratio is large.

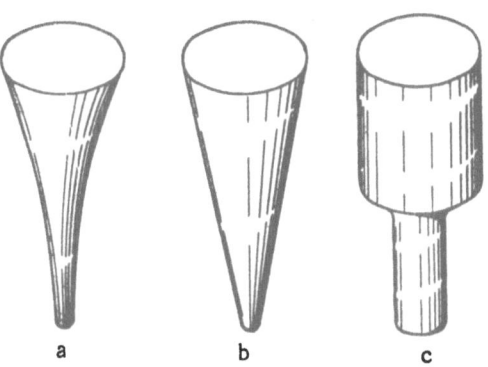

Fig. 74. Various types of circular concentrator:
a) exponential; b) straight conical; c) stepped.

All calculations are at present made on the assumption that plane waves are propagated. This is so only if the transverse linear dimensions are less than half a wavelength; ones of larger size carry more complicated types of wave, which reduces the effect.

Even if the transverse dimensions are less than half a wavelength, they may be comparable with the length, in which case it may be necessary to correct for the presence of radial waves resulting from the Poisson effect. The method of making the correction has been described [18]. For instance, the corrected formula for the length of a circular exponential rod is

$$L = (1 + \Delta)\frac{mc}{f}\sqrt{1 + \left(\frac{\ln N}{2\pi m}\right)^2},\qquad(3.46)$$

in which

$$\Delta = \frac{m^2\pi^2\mu^2}{8}\left(\frac{D}{L}\right)^2\frac{N^2-1}{N^2\ln N},\qquad(3.47)$$

and μ is Poisson's ratio, D being the diameter at the input end.

Similar corrections are readily deduced for concentrators of any shape.

Now we consider the critical frequency. The expression for the phase velocity

$$c' = \frac{c}{\sqrt{1 - \frac{\alpha^2 c^2}{4\omega^2}}}$$

implies that no propagation occurs for frequencies below

$$\omega_c = \frac{\alpha c}{2}.\qquad(3.48)$$

A more detailed consideration shows that for low ω/ω_c even perfect matching of the load leads to the reflection of much of the energy from the concentrator, with the result that relatively little reaches the load.

It is found to be sufficient for the frequency to exceed the critical value by a factor 1.4-1.5 for an exponential part joined to a cylindrical or prismatic one:

$$2\pi f \geqslant 1.5\omega_c.\qquad(3.49)$$

This check against the critical frequency must be made at the end of the calculation in every case apart from that of the stepped rod, for which the critical frequency cannot be calculated if the waves are assumed to be ideally plane. However, there are many instances in which stepped rods designed for very high gain (100 or so) have failed on test to give any appreciable output; clearly, increase in the area ratio N gives increased gain only within certain limits.

67

TABLE 13. Formulas for Types of Concentrator

Type	Law of area	Resonance condition	Gain in amplitude
Exponential	$S = S_0 e^{-\alpha x}$, $\alpha = \dfrac{2\ln N}{L}$	$\sin\left(2\pi\dfrac{f}{c}\sqrt{1 - \dfrac{\alpha^2 c^2}{4\omega^2}}\,L\right) = 0$	$k = N$
Conical	$S = S_0(1-\alpha x)^2$, $\alpha = \dfrac{N-1}{NL}$	$\dfrac{2\pi fL}{c}\left[\cot\dfrac{2\pi fL}{c} - \left(\dfrac{N}{1-N}\right)^2\dfrac{2\pi fL}{c}\right] = 1$	$k = \left\| N\left(\cos\dfrac{2\pi fL}{c} - \dfrac{c}{2\pi fL}\dfrac{N-1}{N}\sin\dfrac{2\pi fL}{c}\right)\right\|$
Catenary	$S = S_0\,\text{ch}^2\,\gamma(L-x)$, $\gamma = \dfrac{\text{arch } N}{L}$	$k_1'L\tan k_1'L = -\sqrt{1 - \dfrac{1}{N^2}\text{arch } N}$; $k_1' = \sqrt{\left(\dfrac{2\pi f}{c}\right)^2 - \left(\dfrac{\text{arch } N}{L}\right)^2}$	$k = \left\|\dfrac{N}{\cos k_1'L}\right\|$
Stepped (symmetrical)	for $\ 0 \leqslant x \leqslant \dfrac{L}{2}\ \ S = S_0$, for $\ \dfrac{L}{2} \leqslant x \leqslant L\ \ S = S_l$, $S_0/S_l = N^2$	$L = \dfrac{c}{2f}$	$k = N^2$
Stepped (unsymmetrical)	for $\ 0 \leqslant x \leqslant L_1\ \ S = S_0$, for $\ L_1 \leqslant x \leqslant L_1 + L_2\ \ S = S_l$, $S_0/S_l = N^2$	$\tan\dfrac{\omega L_2}{c} = -N^2\tan\dfrac{\omega L_1}{c}$	$k = N^2\dfrac{\sin\dfrac{2\pi fL_1}{c}}{\sin\dfrac{2\pi fL_2}{c}}$

In addition, mechanical losses occur in the concentrator, and these sometimes become very important. For instance, a choice of excessive gain may result in impermissibly large deformation, with the result that the concentrator becomes very hot and may fracture. To guard against this, the system must be checked to ensure that at no point does the stress or deformation exceed the limit set by the elastic limit, the yield points, or fatigue limits. The stress distribution is easily calculated for any type of concentrator, so this test represents no special difficulty.

It is rather more difficult to deduce the contribution to the mechanical loss in the concentrator to the overall energy balance; a general calculation requires new terms in the wave equations and becomes extremely complicated even for linear losses (ones proportional to the deformation rate). Some simplified treatments are therefore essential. For instance, the resistive losses in concentrator and load may be small relative to the reactive energy flow (small traveling-wave ratio; see section 15), in which case the method given in section 12 may be used. Here one measures (say) Q and calculates

$$I_v = \int_0^l v^2(x)\, S(x)\, dx \tag{3.50}$$

and

$$I_\sigma = \int_0^l \sigma^2(x)\, S(x)\, dx, \tag{3.51}$$

whence via (3.10), (3.2), or (3.11) the loss factor θ, the loss energy W, or the loss resistance R may be calculated in convenient form.

Interesting results are obtained when different types of concentrator made of the same material (having the same θ) are compared. The linear treatment (section 12) shows that the Q is independent of the gain for the stepped type:

$$Q_s = \frac{E}{\omega\theta}. \tag{3.52}$$

This implies that $Q_s = Q_r$, the latter being that for a uniform half-wave cylinder.

The analogous calculation for a half-wave exponential rod gives

$$Q_e = Q_r \left\{ 1 + \frac{(\ln k)^2}{\pi^2 + (\ln k)^2} + \frac{(\ln k)^4}{[\pi^2 + (\ln k)^2]^2} \right\}. \tag{3.53}$$

The additional factor is a function of the gain and lies within the range

$$1 \leqslant \left\{ 1 + \frac{(\ln k)^2}{\pi^2 + (\ln k)^2} + \frac{(\ln k)^4}{[\pi^2 + (\ln k)^2]^2} \right\} \leqslant 3. \tag{3.54}$$

These results show that exponential concentrators give Q higher by factors of 1.3-1.7 (related to stepped rods); the results are reliable, for they apply to gains in the range 6 to 20.

15. Effects of Load

We have seen in Chapter 2 that the cutting takes only a very small fraction of the input energy of the system as a whole. From this we might conclude that the effects of the load on the vibrating system are unimportant and so need not be discussed; but this is not the case.

Firstly, it is very likely that the efficiency of ultrasonic machines will be raised. It is very likely that current trends will lead to the much greater importance of matching the source to the load.

Secondly, the resistive load has a very important effect on the pattern of traveling and standing waves and hence on the loss of energy through supports.

It is known from radio engineering that the placing of isolators along a long line becomes more critical as the traveling-wave ratio falls; a pure standing wave allows of attachment only at nodes, whereas a pure traveling wave allows of attachment at any point. The same applies to acoustic systems, which have to be fastened to the fixed parts, e.g., via quarter-wave supports. The inevitable errors in choosing points of support and the finite thickness of the parts can result in very substantial losses, especially when the traveling-wave ratio is small.

Of course, the reactive component of the load is also important; this represents the mass of the tool (which varies on account of wear) and some equivalent mass of the workpiece. This component varies, on account of wear or change in the static force, and so the resonance frequency of the whole system is affected. Even manual or automatic tuning will not overcome the resulting displacement of the nodes, while detuning will alter the node pattern, coupling between parts, and so on, which may increase the loss or otherwise diminish the useful effect. The effects of change in the wave phases are the less the greater the traveling-wave ratio.

The type of concentrator has an important bearing on the effects of load parameters on the system as a whole [20, 21]. As an example we compare half-wave stepped and exponential types, both working into a specific mechanical impedance z_0, which in this case we take to be a pure resistance r_0.

The traveling-wave ratio T for the exponential type is found from (3.37)-(3.39). The boundary condition

$$\frac{\sigma}{v} = r_0 ,\tag{3.55}$$

with x = c'/2f gives the reflection coefficient as

$$\beta = \frac{-\frac{\alpha}{2} + j\frac{\omega}{c}\left(\sqrt{1+\left(\frac{\ln N}{\pi}\right)^2} - \frac{\rho c}{r_0}\right)}{\frac{\alpha}{2} + j\frac{\omega}{c}\left(\sqrt{1+\left(\frac{\ln N}{\pi}\right)^2} + \frac{\rho c}{r_0}\right)}\tag{3.56}$$

Line theory gives

$$T = \frac{1-|\beta|}{1+|\beta|} .\tag{3.57}$$

Figure 75 (full lines) shows T for the exponential type, whose parameter is the area ratio N (which equals the gain). The load factor is

$$\gamma = \frac{r_0}{\rho c} .\tag{3.58}$$

The expressions for the stepped type are rather more cumbrous but essentially similar; the reflection coefficient at the input end is

$$\beta = \frac{1 - \frac{\rho c}{r_0}\frac{1}{N^2}}{1 + \frac{\rho c}{r_0}\frac{1}{N^2}} ,\tag{3.59}$$

whence from (3.57) we readily find T. The broken lines in Fig. 75 give results for this case, in which N, or the corresponding gain k, is the parameter. The actual load factors lie between 0 and 1, and the curves show that T is always substantially larger for the exponential type in this range for equal k as well as for equal N.

The effects of a reactive or complex load can also be examined by reference to (3.37)-(3.39) and the corresponding equations for the stepped type. The boundary condition at x = L is

$$\left(\frac{\sigma}{v}\right)_{x=L} = z_0 ,\tag{3.60}$$

whence the input impedance is *

$$z_i = S_0 z_0 = S_0\left(\frac{\sigma}{v}\right)_{x=0} .\tag{3.61}$$

------

* Here and subsequently we use the method of averaging the acoustic parameters over the cross section that is employed in the deduction of Webster's equation (section 14).

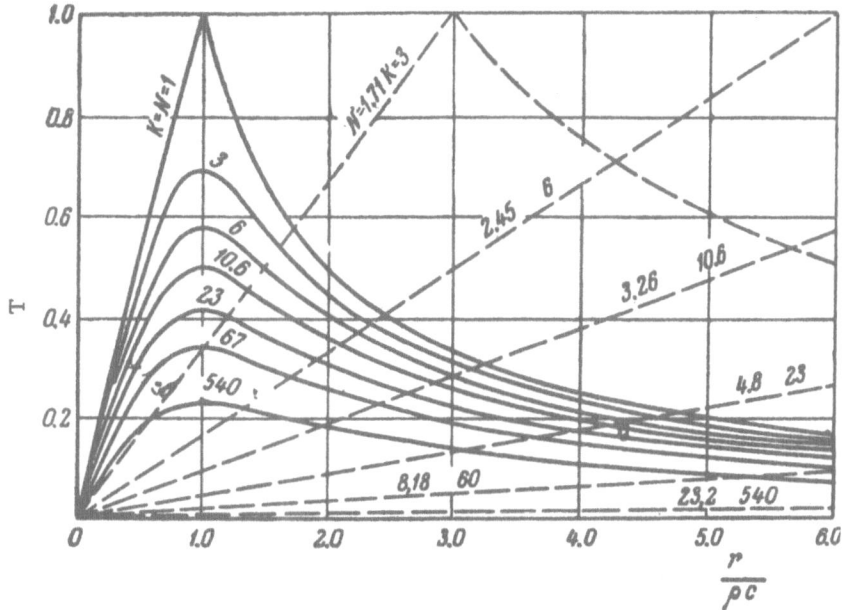

Fig. 75. Load factor and traveling-wave ratio for exponential and stepped concentrators.

We now assume that this is uniformly distributed over the exit area, being

$$z' = S_1 z_0 \ , \tag{3.62}$$

in which case we have

$$z_{is} = N^2 z' \ ; \quad z_{ie} = N^4 z'. \tag{3.63}$$

This means that the load impedance is referred from exit to input by multiplication by the square of the area ratio, whereas the fourth power is required for the stepped type.

The deductions to be drawn are that the exponential type is better as regards matching and detuning effects, although the gain is somewhat less and the shape is rather more difficult to produce. The final choice must be made in accordance with the purpose of the machine.

For instance, the machine may be of universal type; the various types of tool must then be attached to the end of the concentrator, so a stepped rod is preferable. On the other hand, the machine may be meant for some special operation on hard alloys, in which case the exponential type is better.

Again, the stepped form is to be preferred in machines for working relatively soft materials such as glass and semiconductors; this is especially so if the tool is to be made integral with the concentrator.

The following simple method allows one to examine the detuning effects of interchangeable tools attached to the end of the concentrator.

The tool has a mass M_0, which may be determined by weighing. The relative change in the resonance frequency is

$$\nu_0 = \frac{\Delta f_0}{f} \ .$$

Then the change produced by another tool of mass M is given by the approximate formula

$$\nu = \frac{\Delta f}{f} = \frac{1}{\sqrt{1 - \dfrac{M}{M_0}\left[1 - \dfrac{1}{(1+\nu_0)^2}\right]}} \ . \tag{3.64}$$

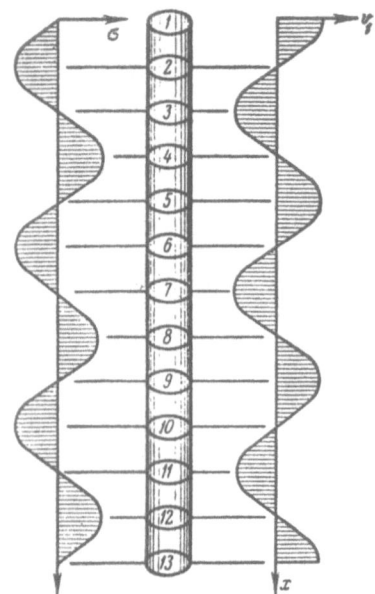

Fig. 76. Standing waves in a rod system.

A permissible range of tool masses should be specified for a given machine, which is possible if we know ν', the maximal change in frequency without excessive increase in loss or undue change in source adjustment; then from previous measurement we have the maximal mass of the tool as

$$M' = \frac{(1-\nu')^2}{1-\dfrac{1}{(1+\nu_0)^2}} M_0. \qquad (3.65)$$

In using (3.64) and (3.65) it is to be remembered that they involve the assumption that the tool is very small relative to the wavelength (ratio 20-30) and that ν_0 and ν' are always negative, since the addition of a localized mass reduces the resonance frequency.

16. Selection and Design of Demountable Joints

The main such joints in the acoustic part are butt joints and joints between the acoustic head and the fixed parts.

The former join parts of the main wave path; they are to allow replacement or interchange of tools and concentrators, or parts of these. In some cases the tool is integral with the concentrator, or may be welded or soldered to it. The joint may be formed by a screw, by a clamping nut, by a taper shank, or in other ways.

Any demountable joint is bound to have some gaps, so the wave path is divided into separate, not very closely linked parts. If the joint lies at a point where calculation would indicate a node, the system becomes divided into separate resonant elements tuned to the same frequency, which means that good transfer occurs even when the coupling is weak. On the other hand, a joint at any other point gives parts tuned to different frequencies, and the greater the frequency difference the poorer the matching to the load, so the lower the efficiency.

This means that a rod as of Fig. 76 should have its joints at the points shown by the odd numbers, or near these. In the simplest case (a system of length two half-wavelengths), there is a half-wave transducer and a half-wave concentrator, so the only two possibilities are an integral concentrator and tool or a small replaceable tool bit. More complex designs allow of a joint at some intermediate point where there is a pressure node.

A very different approach is needed for the attachment of the head to the fixed parts, for the resulting shunt should have the minimum possible admittance in order to minimize the loss of energy from the head.

The earliest designs employed attachment in antinodal planes (those shown in Fig. 76 by even numbers), usually by means of collars in the central plane of the core or in other similar ways. There are two suitable planes in a two half-wave system, one at the middle of the transducer and the other somewhere along the concentrator rather nearer the broad end than the working end. A more complex system has as many antinodes as there are half-wavelengths.

Any error in location of the nodes directly affects the performance as regards loss of energy to the fixed parts. Moreover, incorrect attachment introduces additional reactance into the system and so affects the tuning. The nodal plane of a symmetrical transducer lies exactly at the center; a symmetrical stepped transducer for a tapering concentrator if the equation is known for the velocity or displacement. This calculation gives [17] the following results for the various types of concentrator:

Exponential

$$\tan \frac{\omega x_0}{c'} = \frac{2\omega}{\alpha c'}, \qquad (3.66)$$

Conical

$$\tan \frac{\omega x_0}{c} = \frac{\omega}{\alpha c}, \tag{3.67}$$

Catenary

$$\tan k_1^1 x_0 = \frac{k_1^1}{\gamma \, \text{th} \gamma l}, \tag{3.68}$$

in which x_0 is the distance from the exit end to the plane; the other symbols are as in Table 13.

Attachment in the nodal plane has the advantages of simple design, relative ease of location, and so on; but it sometimes introduces difficulties. For instance, the planes in a two half-wave system both lie at very inconvenient points; the middle of the transducer does not provide sufficient rigidity, makes it difficult to mount the coils, and interferes with the cooling, while the middle of the concentrator makes it impracticable to use replaceable concentrators.

These difficulties led to the development of other systems, in which the support is given at some other point via a part in which standing waves are generated. The shape and size of this part are made such that the point of attachment to the body of the machine lies at a displacement node in the standing wave. Such supports are wave isolators.

Supports operating with longitudinal waves (Fig. 77) are now commonly used; they have the advantages of simplicity and high mechanical rigidity. They usually operate at an antinode, and the walls are usually made thin; the axial length is a quarter wavelength, the point of attachment ɔ the fixed parts being a heavy flange. The joint then lies at a displacement node, which gives rise to strong ɾ ɭection, on account of the change in acoustic impedance. Such an isolator may be mounted either way up, ω enclose the transducer or concentrator [22, 23].

Such an isolator need not be a quarter-wave one, for use is made of the phase relationships in the standing waves; it can be attached at any point, but in that case the phases must be such that the outer edge is always a displacement node. This feature makes the design problem much easier. Figure 78 illustrates the use of complementary phases; the external edge of the support is a displacement node, although the inner edge does not lie at an antinode. The residual loss through the support can be minimized by making the wall thickness variable (Fig. 79); the isolator then acts as a reversed concentrator (reduces the amplitude). As in the concentrator (section 14), we have here a phase velocity distinct from that for a uniform rod of the same material, so the geometrical length is greater than the distance between node and antinode in the main path.

Disc isolators operating in bending (Fig. 80) have recently come into use; the thickness and diameter are made such that the outer edge is a displacement node for the bending vibrations at that frequency when the inner edge is excited. Here we must distinguish two modes of support of the outer edge, namely clamped (Fig. 81a) and hinged (Fig. 81b).

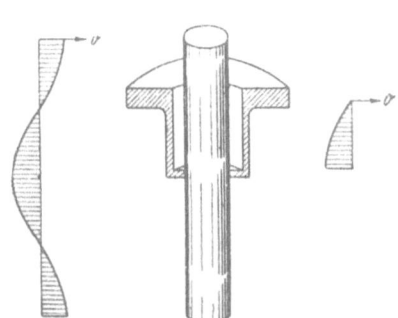

Fig. 77. Quarter-wave isolator (support).

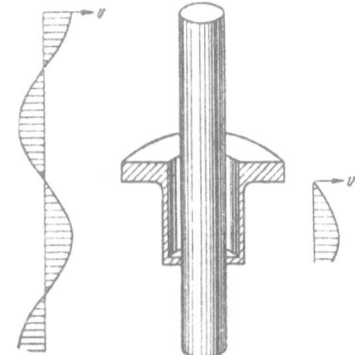

Fig. 78. Choice of point of attachment of quarter-wave support for any wavelength.

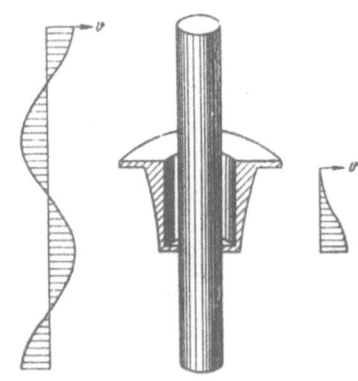
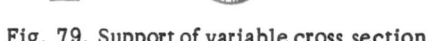

Fig. 79. Support of variable cross section.

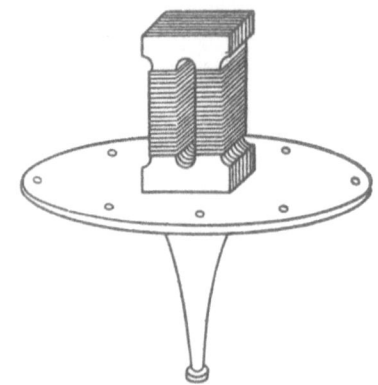

Fig. 80. Disc support.

The fundamental of a circular plate loaded at the center and clamped at the edge is [24]

$$f_0 = 1.88 \frac{h}{D^2} \sqrt{\frac{E}{\rho(1-\mu^2)}} \, , \qquad (3.69)$$

in which h is thickness (cm), D is diameter (cm), E is Young's modulus (dyne/cm²), ρ is density (g/cm³), and μ is Poisson's ratio. Reducing the actual distribution in the perturbing force to a force localized at the center, we have the outside diameter as

$$D = 1.37 \sqrt[4]{\frac{h^2}{f^2} \frac{E}{\rho(1-\mu^2)}} \, . \qquad (3.70)$$

Table 14 gives the parameters in this formula for some actual materials.

For instance, a steel disc has to operate at 20 kc; even for the fairly large thickness h = 5 mm = 0.5 cm we have

$$D = 1.37 \sqrt[4]{\frac{0.5^2}{4 \cdot 10^8} \frac{2 \cdot 10^{12}}{7.8(1-0.28^2)}} = 4.98 \cong 5 \, \text{cm}.$$

This is applicable only if the thickness is not too large in relation to D, otherwise the real distribution of the force will differ too greatly from that assumed in the calculation.

Hinged discs (Fig. 81b) allow of relatively free angular displacement in planes perpendicular to the line of external support; such oscillation gives rise to comparatively little reaction. The same effect is produced by machining a circular groove of depth 0.6-0.8 of the thickness, as shown in Fig. 81c.

The fundamental for this case, in terms of the same symbols, is [24] given by

$$f_0 = 0.9 \frac{h}{D^2} \sqrt{\frac{E}{\rho(1-\mu^2)}} \, , \qquad (3.71)$$

TABLE 14. Parameters of Some Materials

Material	E, dyne/cm²	ρ, g/cm³	μ
Aluminum	$0.7 \cdot 10^{12}$	2.70	0.33
Brass	$1.0 \cdot 10^{12}$	8.50	0.37
Nickel	$2.04 \cdot 10^{12}$	8.86	0.31
Steel	$2.0 \cdot 10^{12}$	7.80	0.28

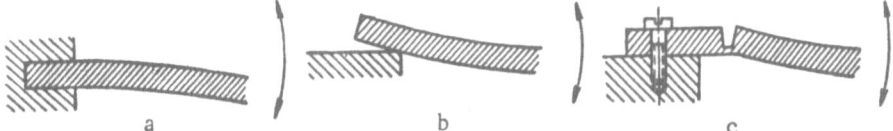

Fig. 81. Methods of fixing the edge of a disc support.

and so

$$D = 0.95 \sqrt[4]{\frac{h^2}{f^2} \frac{E}{\rho(1-\mu^2)}} \, , \qquad (3.72)$$

which is less than before, other things being equal, so this method can be used only for low-power machines with small components. On the other hand, the groove has a major advantage; it introduces a sudden change in the acoustic parameters and so causes strong reflection, which minimizes the loss to the fixed parts.

The disc isolator is of much greater value if used in overtone modes (that is, when the external edge is not the first nodal line). The natural frequencies of a plate clamped at the edge are

$$f_0, \ 3.88 f_0, \ 8.8 f_0 \, , \text{etc.}, \qquad (3.73)$$

whereas those for a hinged plate are

$$f_0, \ 6 f_0, \ 15 f_0, \ 28 f_0 \ , \text{etc.} \qquad (3.74)$$

Extraction of the square root shows that the grooved disc gives more widely spaced nodes and antinodes; the wave pattern along the radius varies more smoothly.

This means that overtone operation enables one to locate the actual point of support far from the center of the disc, which is sometimes of value. In addition, the actual oscillating system can be of fairly large diameter, provided that the inner edge lies at or near an antinode (it is sufficient for practical purposes for it to lie half-way between adjacent nodes), in which case the calculation can be performed as though the disc were solid (had no hole), though the diameter of the antinodal line is then fixed. The calculation can be performed via (3.69), (3.70), or (3.72) with the appropriate coefficients from (3.73) or (3.74). The hinged style is preferable, since it gives a smoother wave pattern with more widely spaced antinodes and so makes the calculation less critical; in addition, the outside diameter is larger.

We conclude with a few general comments on disc isolators. The thickness should always be minimized, for a thick isolator causes excessive transfer to the fixed parts not only via bending modes but also via shear and longitudinal waves (on account of Poisson's ratio).

The minimal thickness is governed by considerations of strength in relation to diameter. It is undesirable to use very high overtones, for this tends to increase the mechanical loss in the isolator. Another means of minimizing the loss is to make the isolator of radially increasing thickness (Fig. 82), which reduces the amplitude of the traveling wave at the edge and also increases the distance between nodes; but calculations become rather difficult. Alternatively, the groove at the edge can be replaced by a pressed rim (Fig. 83).

Combined isolators (ones operating in bending and longitudinal compression; Fig. 84) are of interest. Here the problem is to combine the bending in the disc with longitudinal oscillation in the vertical part to give a node at the outside edge [25, 26].

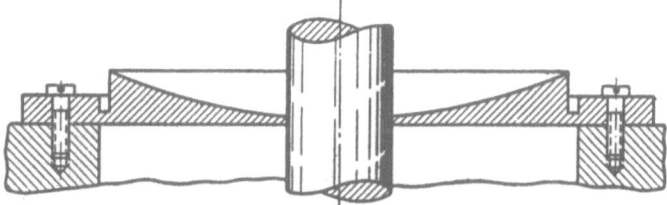

Fig. 82. Disc isolator of variable thickness.

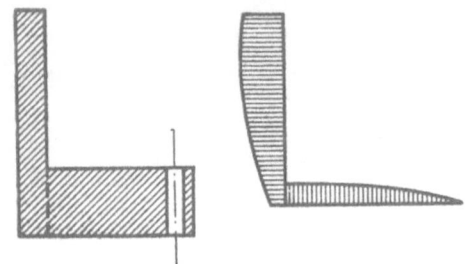

Fig. 83. Attachment
with a pressed rim.

Fig. 84. Support with bending and compressional
modes (use of phase complement).

17. Some Types of Acoustic Head

Certain requirements are imposed on the head in any specialized machine; these have led to the introduction of new types of vibrating system, which differ in several ways from the basic type discussed above.

There is often a need for small heads, as in portable or low-power machines (for working materials such as ceramics and ferrites, engraving, drilling), multihead machines, and so on. Here we discuss briefly some ideas for miniature heads.

Figure 85 shows one method of miniaturization, in which the magnetostriction vibrator is also the concentrator [27]. This roughly halves the length and saves the metal that would form the concentrator. The disadvantages are that the magnetic circuit has gaps and that the lateral dimensions are increased; moreover, it becomes difficult to attach the vibrating system to the fixed parts and to provide water cooling. Finally, the amplitude in the laminated material becomes large, which is most important of all, because such materials show very much more mechanical loss than continuous solids do; further, the mechanical losses of magnetostrictive materials are much higher than those of ordinary metals. For instance, nickel cannot be used, for its elastic limit is exceeded at amplitudes above 3-5 μ, with consequent rapid increase in loss.

Figure 86 shows another method, in which a quarter-wave vibrator is combined with a quarter-wave concentrator [28]. Here there are no gaps in the magnetic circuit, which simplifies clamping and cooling problems. Such a system, if mounted on a plate, has very high transverse rigidity, which is important for precision machining. The region of large amplitudes is also transferred from the vibrator to the concentrator.

The disadvantages here are the relatively small volume of magnetostrictive material and the need for a joint at the region of maximal stress. The first restricts the power, while the second means that the only demountable joint can be at the tip of the concentrator. The vibrator must be hard-soldered to the concentrator.

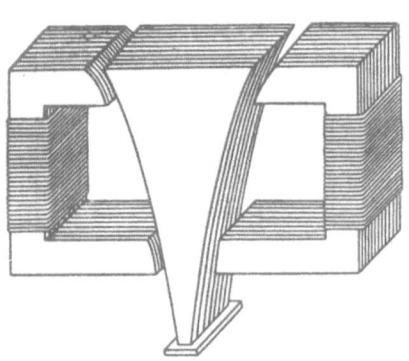

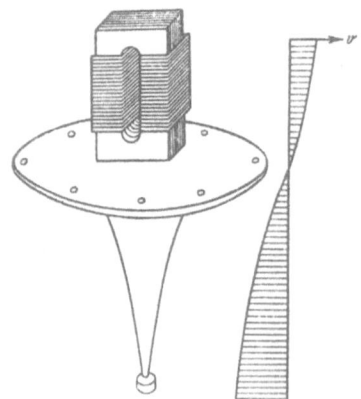

Fig. 85. Combined vibrator and concentrator.

Fig. 86. Oscillating system,
half wavelength long.

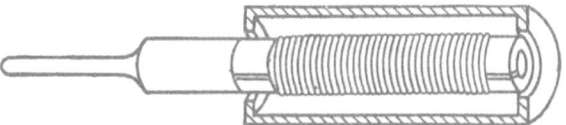

Fig. 87. Acoustic head with tubular transducer.

Dentists' drills and hand engraving machines require very light and compact heads; in this case it becomes impracticable to use laminated materials, so simple rods or tubes of magnetostrictive material are used. Figure 87 shows one such style.

A tube is preferable to a rod, since the cavity can be used for water cooling, which is especially important, since the magnetic losses are very much higher. Some reduction in the eddy-current loss is possible if the tube is slit along its length. The magnetic impedance is reduced, and the magnetic circuit is closed, by making the body of magnetic metal. The materials that commonly have to be cut (e.g., hard parts of teeth) are relatively soft and do not cause much wear on the tool, so it is usually acceptable to employ stepped concentrators. The cutting bit is attached to the end of the concentrator.

An even more promising system, which satisfies much the same requirements, has been made as follows: a thin strip of magnetostrictive material is coated on one side with an insulating film. This is wound on a long rod, which is afterwards removed. The resulting laminated tube has all the valuable properties of an ordinary tube but has lower magnetic losses, so it is more efficient and needs less cooling. Further, no slit along the length is needed.

The system with bent waveguides is of particular interest; the problem arose in connection with the first ultrasonic dentists' drills, for the cutting tool needs to be fixed to the end of a bent rod in order to give convenient access to the tooth. The concentrator is made with a curved end, the tip being the tool. Provided that the radius of curvature is small relative to the wavelength, there is little effect on the cutting; in particular, sharp bends through 90° are accessible.

Fig. 88. Machine with four bent waveguides.

This feature makes it possible to design machines with many cutting heads; Figure 88 shows an American machine (by Sheffield) of this type. The acoustic head contains only one vibrator (not shown), to which is joined a stepped concentrator with four long outgoing waveguides, which turn through a right angle to feed four cutting heads working on four workpieces attached to independent tables [29].

This system has many advantages over ones employing independent acoustic heads. Firstly, lack of agreement between resonance frequencies makes it difficult to feed several heads from one amplifier. Secondly, a machine to perform several operations simultaneously would need several miniature heads, which, though possible, involves increased mechanical and magnetic losses. The use of a single laminated transducer provides higher efficiency and simplifies the cooling.

Another interesting innovation is ultrasonic milling, in which a groove is cut by moving the workpiece in a plane perpendicular to the plane of vibration. The commonest method of doing this is to use a knife-type concentrator (Fig. 89) with a leading edge somewhat rounded. This method works, but the cutting rate is poor and the vibrating system is subject to a constant lateral thrust, which tends to excite bending vibrations and so causes an increase in the mechanical losses.

Figure 90 shows a Japanese invention [30] designed to overcome this, in which a cylindrical transducer operating with low-frequency radial vibrations is attached to a disc concentrator tapering towards the edge, the thickness at the edge being the width of the slot to be milled. This edge is also the cutting tool. The disc rotates about its axis to move the cutting point along the workpiece.

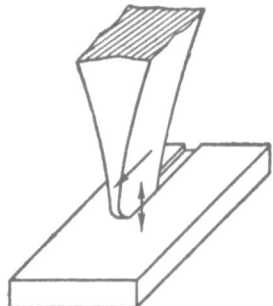

Fig. 89. Ultrasonic milling.

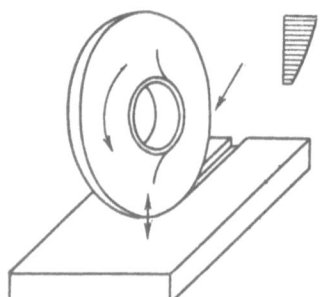

Fig. 90. Disc ultrasonic milling cutter with ring vibrator.

As yet there is no reliable information on the practical use of this interesting scheme, but the idea of a disc system may prove useful in future developments.

In some cases, when very hard materials have to be cut, it is more convenient to interchange tool and workpiece; the latter is attached to the end of the concentrator, while the tool is fixed.

To conclude, we note that some recent papers [31-33] relate the parameters of acoustic heads to the stress distribution and strength of the materials. Although all of these deal solely with concentrators, the main ideas can be turned to advantage in relation to other parts of the heads.

Literature Cited

1. L. Bergmann, Ultrasonics [Russian translation], IL, 1956.
2. I. I. Teumin, Ultrasonic Vibratory Systems, Moscow, Mashgiz, 1959.
3. Z. N. Bulycheva, E. I. Gurvich, and Ya. P. Selisskii, "Magnetic alloys used in ultrasonics," in: Industrial Use of Ultrasonics, Moscow, Mashgiz, 1959, p. 91.
4. K. P. Belov, Elastic, Thermal, and Electrical Effects in Ferromagnetics, Moscow, GITTL, 1957.
5. L. Ya. Gutin, "Theory of the magnetostriction transducer," Zh. Tekhn. Fiz. 15(4-5):239, 1945.
6. L. Ya. Gutin, "Magnetostriction radiators and detectors. I. Vibrator of rod type," Zh. Tekhn. Fiz. 15(12): 924, 1945.
7. R. Suwalski, Obliczenie elektrycznego ukladu zastepczego obciqzonege pretowego przetwornika magnetostrykcyjnego," Arch. elektrotechniki (ПРН) 7(4):647, 1958.

8. G. V. Glebovich, "Calculation of vibrators of industrial ultrasonic machines," Tr. Gor'kovsk. Politekhn. Inst. 13(1):30, 1957.

9. L. O. Makarov, "Design of an instrument for mechanical working of hard materials," Report, Acoustics Institute, Academy of Sciences of the USSR, Moscow, 1955.

10. A. Nomoto, "Ultrasonic machining by low power vibration," J. Acoust. Soc. Am. 26(6):1081, 1956.

11. M. E. Weiss, Ultrasonic drill, US Patent No. 2831295, dated Sept. 29, 1955.

12. C. M. van der Burgt, "Piezomagnetic ferrites," Electronic Technology 37:330, Sept. 1960.

13. W. P. Mason, Method of obtaining velocity with crystals, US Patent No. 2514080, dated Jan. 10, 1945.

14. W. P. Mason and R. F. Wick, Mechanical impedance transformer, US Patent No. 2573168, dated May 23, 1950.

15. M. G. Lozinskii and L. D. Rozenberg, Method of concentrating ultrasonic energy, Authors' certificate, USSR No. 85193, August 4, 1949.

16. N. N. Andreev, "The averaging method in the solution of wave problems," in: 70th Anniversary of Academician A. F. Ioffe, Moscow, Izd. Akad. Nauk SSSR, 1950, p. 467.

17. L. G. Merkulov, "Theory of ultrasonic concentrators," Akust. Zh. 3(3):230, 1957.

18. M. M. Pisarevskii, "Calculation of intermediate rods for magnetostriction vibrators," in: Calculation of Magnetostriction Transducers, Izd. Mosk. doma nauchno-tekh. propagandy im F. É. Dzerzhinskogo, 1957, pp. 29-41.

19. L. O. Makarov, "Method of calculating rod-type exponential ultrasonic concentrators," in: Industrial Use of Ultrasonics, Moscow, Mashgiz, 1959, p. 102.

20. L. O. Makarov, "Waveguide features of rod-type ultrasonic concentrators," in: Use of Ultrasonics in Quality Control, Testing, and Machining of Metals and Alloys, Kiev, Izd. Akad. Nauk UkrSSR, 1960, p. 44.

21. J. Kacprowski, "Analiza parametrow falowych tuby wykladniczej," Arch. elektrotechniki (ПHP) 5(4):719, 1956.

22. C. L. Calosi, Support for vibratory devices, US Patent No. 2632858, dated March 23, 1953.

23. D. F. Yakhimovich, Head for ultrasonic machining, Author's certificate, USSR No. 109727, dated August 10, 1956.

24. V. K. Iofe and A. A. Yanpul'skii, Graphs and Tables for Electrical-Acoustical Calculation, Moscow-Leningrad, Gosenergoizdat, 1954, p. 93.

25. P. V. Kobin, An ultrasonic radiator, Author's certificate, USSR No. 113357, dated Feb. 9, 1957.

26. C. F. Brockeslby, "High power transducers for frequencies around 20 kc/s, "Commun. congr. internat. sur les traitements par les ultra-sons, Marseille, 1955, p. 113.

27. M. M. Pisarevskii, "Ultrasonic method of machining hard materials," Stanki i Instr. No. 5:16, 1954.

28. V. I. Fomin, L. O. Makarov, and D. F. Yakhimovich, An ultrasonic vibrator, Authors' certificate, USSR No. 121606, dated Dec. 16, 1956.

29. "New tool shoots ultrasonic energy around corners," Ultrasonic News, No. 1:16, 1959.

30. Fukumota, Machinery (Japan) 22(8):1281, 1959.

31. C. Kleesattel, "Vibrator ampullaceus," Acustica 12(5):322, 1962.

32. E. A. Neppiras, "Mechanical transformers for producing very large motion," Acustica 13(5):368, 1963.

33. E. Eisner and J. S. Seager, "A longitudinally resonant stub for vibrations of large amplitude," SMRE Research Report No. 216, Oct. 1963, 51 pp.

CHAPTER 4

ULTRASONIC MACHINE TOOLS

18. Kinematic Schemes for Ultrasonic Machining

The oscillatory motion of the tool may be combined with a lateral or angular feed to perform a great variety of ultrasonic cutting operations [1, 2].

Many ultrasonic broaching operations can be performed, as in the making of holes of complex profile, recesses, trepanning, and slicing (Fig. 91a-d). Further, we have ultrasonic drilling, in which the tool rotates (Fig. 91e), which substantially increases the accuracy and reduces the chances of cavitation pitting, though only shapes of rotational symmetry can be formed in this way, of course.

A hole with a curved axis can be cut if the feed direction is varied suitably (on a radius in Fig. 91f). The main difficulty here is to transmit the vibrations to the tool, for curvature in the tool tends to excite undesirable transverse oscillations.

Ultrasonic machines resemble vertical milling and drilling machines in many features, especially in the ways for controlling the relative motion of tool and workpiece. *

Horizontal motions are usually applied to the workpiece, and vertical ones to the tool (sometimes to the workpiece also). Rotating heads or tables are employed to facilitate the replacement of interchangeable tools and the setting of workpieces. The feed is often applied to the tool.

It is common for the tool to break when the holes are less than 0.3 mm in diameter; moreover, the large mass of the head introduces difficulties in feed control for these small sizes. In such cases the vibrational motion and feed are applied to the workpiece instead of the tool. The feed is also applied to the workpiece in

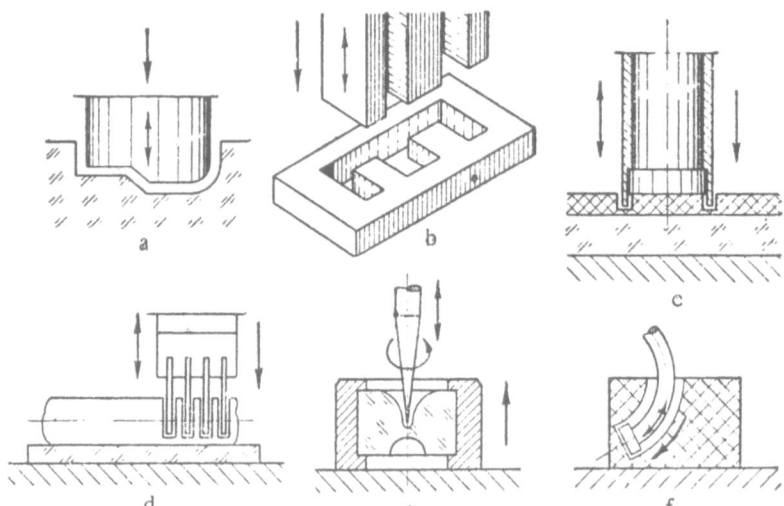

Fig. 91. Ultrasonic broaching: a) recess of special profile; b) profile hole; c) trepanning; d) slicing; e) drilling; f) boring of hole with curved axis.

* Horizontal machines have recently been introduced in the USSR and elsewhere; these provide rapid removal of the used abrasive suspension [3].

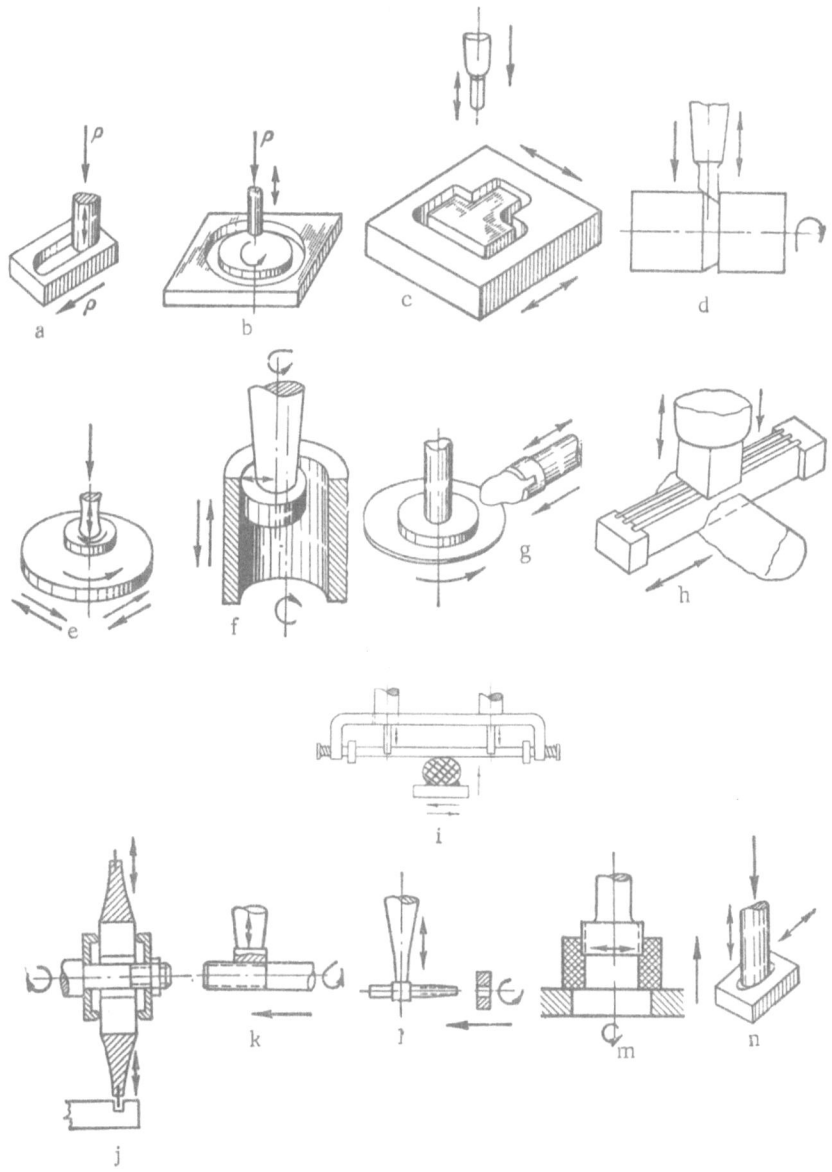

Fig. 92. Various operations in ultrasonic machining: a) slot cutting; b) cutting large discs; c) machining along a complex profile; d) turning; e) surface grinding; f) honing (tool performs torsional oscillations about the axis of the hole); g) cutting with a rotating disc tool; h) cutting with a strip tool; i) cutting with a head having a transducer disc; j) forming external screw thread; k), l), and m) forming internal screw thread; n) machining an elliptical hole.

some large machines, in which case the forces needed to attach the concentrator are not transmitted to the feed mechanism, since the head is firmly fixed to the frame.

The next sections deal with particular parts of the machines (acoustic heads, feed mechanisms, tables, abrasive-feed systems, and generators).

Some novel methods in ultrasonic cutting are at present under development.

Displacement of the workpiece in a plane perpendicular to the direction of vibration of the tool enables one to perform typical milling operations, such as slotting (Fig. 92a), cutting circular blanks of large size (Fig. 92b) [4], engraving, and profile milling (Fig. 92c). Ultrasonic turning (Fig. 92d) is also possible [5]. Surface grinding is effected by giving the tool and workpiece combined motions in a plane perpendicular to that of vibration (Fig. 92e).

Cylindrical grinding and honing are illustrated by Fig. 92f. Here the disc performs rotational oscillations about its axis, the disc also having rotational and axial motions [6].

Slicing can be performed as for slotting and grinding, but a thin tool is used, sometimes with several blades. A rotating disc is used if the wear is very heavy* (Fig. 92g) [7].

Figure 92h shows [8] slicing by means of multiple blades, which are driven by a detached concentrator; Figure 92i shows the use of a transducer disc for this purpose [9].†

Figures 92j to 92l show internal and external thread cutting; the operation is that of normal machining, with the feed in the direction of vibration. The cutting of an internal thread is more complicated, because the tool must perform transverse vibrations, which must be of defined amplitude in order to keep the system in resonance [10, 11].

If the tool performs simultaneously or in turn both lengthwise and transverse vibrations, it is possible to machine an elliptic hole or one whose diameter increases from entrance to exit (Fig. 92m). This approach is of interest in dentistry [12].

Attempts to use the side faces of a longitudinally vibrating tool in order to perform cutting have proved fruitless, in view of the low cutting rates.

19. Design of Acoustic Heads for Ultrasonic Machine Tools

Only some of the vibrators dealt with in Chapter 3 are at present used in these machines (Table 15, Fig. 93) [13].

Only recently has attachment via a resonant disc isolator come into use; this is a very promising method, especially for machines of low or medium power.

Most current machines employ magnetostriction transducers made of nickel, permendur, and alfer. Nickel is the most used, since the methods of handling this are better developed.

The preference for nickel also arises in part from the high strength of the metal and the good insulating properties of a film of nickel oxide. It is more difficult to produce such films on permendur and alfer; in particular, the oxide film on the latter is of low mechanical strength. Treatment with bakelite lacquer or material such as BF2 adhesive is not advisable, because these prevent subsequent hard-soldering to the cores. Tests are being made on inorganic adhesives.

Nickel transducers are made as follows [14]. Sheet nickel (0.1-0.2 mm) is stamped to give core sections‡ and is flattened between plates. These sections are carefully degreased and are then heated to 850°C in an electric oven for 3 h with free access of air to produce the oxide film. The sections are made up into a bundle, whose end is ground; to this is attached with hard solder (PSr40, GOST 8190-56) the fixed part of the concentrator or a holding flange. This part is usually made of U7 steel, since this takes the hard solder well, is readily heat-treated, and is easily polished. Stainless steel type 4Kh13 is even better; this is also easily heat-treated. The flange and core together are then machined to make the two coaxial (any deviation from this may give rise to undesirable transverse vibrations). Then the flange is hardened and the points of attachment are surface-ground.

---

* It is sometimes preferable to use a strip moving lengthwise rather than a rotating disc.
† A method (Fig. 92k) has also been proposed in which the blades are fixed to one concentrator (or several).
‡ Material up to 0.5 mm thick can be used in the 20 kc range; the amplitude is reduced by not more than 3.5% [15]. The object of lamination is to minimize the heating.

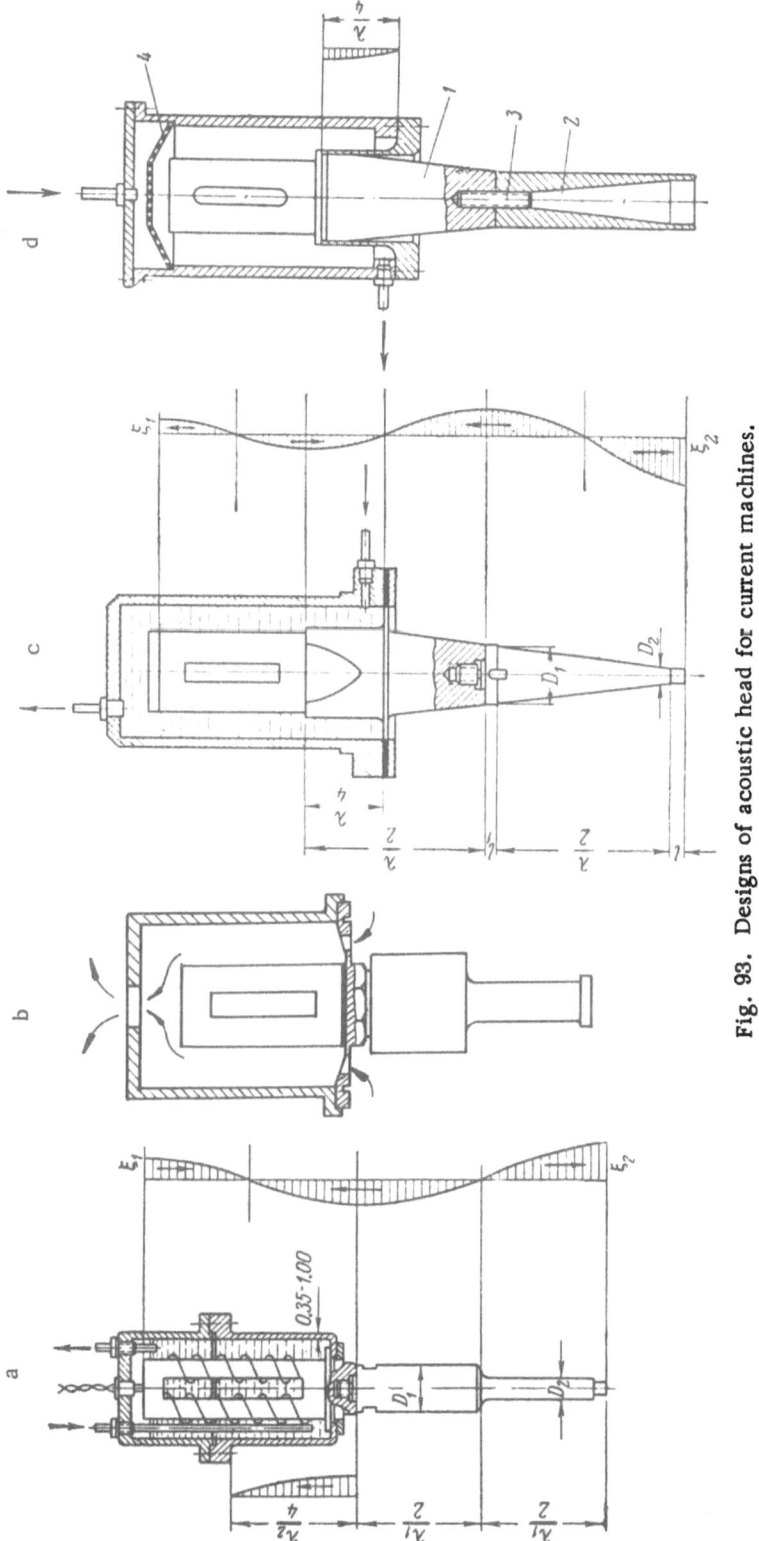

Fig. 93. Designs of acoustic head for current machines.

TABLE 15. Types of Vibrating System Used in Current Ultrasonic Machines

System	Application	Advantages	Disadvantages
Two half-waves with support via a sleeve enclosing the transducer (Fig. 93a)	In machines of power up to 0.5 kW with liquid cooling of transducer	Compactness and rigidity	Complexity, heavy loading of sleeve
Two half-waves with supporting flange (Fig. 93b)	In machines of power up to 0.5 kW with liquid or air cooling of transducer	Compactness, accessibility of core, suitability for air cooling	Heavy loading of flange
Three half-waves with attachment via a flange at the node in the permanent part of the concentrator (Fig. 93c)	In machines of power 1-2.5 kW with water cooling of transducer	Soundness of design	Large length of vibrating system
Three half-waves with attachment via a sleeve enclosing the permanent part of the concentrator (Fig. 93d)	In machines of power 0.5-1 kW with water cooling of transducer	Rigidity	Complexity, heavy loading of sleeve

The transducer is commonly made prismatic with two rods, since this is simple in design and most suitable for air cooling. Single-rod systems (with closure of the magnetic circuit via an additional armature) have rarely been used; the same applies to single-rod open-circuit ones and to three-rod ones.

A low-power transducer may be assembled from thin-walled nickel tubes soldered to the flange. If the tubes vary in length (e.g., as a result of rounding at the free end), the assembly has a broader resonance curve, which simplifies tuning and makes the system less sensitive to wear in the tool [63].

The winding is done with wire having waterproof insulation (such as PMVG wire). The best winding is a single-layer coil for excitation and bias magnetization, but sometimes two separate coils are used in order to simplify the matching to the generator. Current densities up to 20 A/mm^2 can be used with water cooling.

Much of the energy fed to the transducer appears as heat; natural air cooling is used in machines of power up to 50 W. The moving system may be held via disc isolators, because this leaves the transducer open. It is particularly simple to produce a rotating head in this case.

Water cooling is essential at high powers, which introduces some difficulty, especially in view of the need to prevent corrosion and cavitation erosion; chromium plating, oxide coating, and lacquering are quite useless. The best approach is to make the parts of stainless steel (e.g., Kh18N9T), which is most resistant to cavitation erosion. The surfaces liable to erosion may also be coated with rubber [16].

Measures must be taken to suppress the radiation from the free end of the transducer; for this purpose the cooling jacket is extended below the end (Fig. 93a) or rubber sponge is attached to the end (Fig. 93c), the latter being suitable for a vibrator to be used in any position. The vibrator may also be cooled by a water spray (Fig. 93d).

The outer surfaces must be protected from corrosion arising from condensation if cold water is used. This also makes it undesirable to enclose the head within the feed mechanism, but this is not always possible, especially in high-power machines.

The water line should include a relay to trip the supply to the transducer in the event of flow failure.

Water cooling introduces difficulties, and there are advantages in oil cooling; the oil fills the cavities in the head and is itself cooled by air fins or by a coil carrying water. Both methods are effective only for powers of the order of 100-150 W.

The parts (flanges and so on) that hold the head have to operate under heavy load; they soon fail by fatigue if of poor design and manufacture. The surface finish should be good and there should be no sudden changes of section, and especially no notches or scratches. Supporting plates should be not less than 0.5 mm thick, while the radii of curvature at the point where the flange joins the concentrator should be not more than 4 mm [15].

Interchangeable concentrators are attached to the transducer by screwed shanks and locknuts. The latter enable one to set the tool in a definite position relative to the workpiece without turning the head. However, this mode of attachment restricts the diameters of tool and concentrator, and does not always provide a reliable joint.

The concentrator must be made of a material with low acoustic losses and high fatigue resistance; the material must also be one that solders well. Monel metal has all these properties, but the usual constructional steels are nearly as good, so concentrators are nearly always made of these (66S2, 65G, 40Kh, and so on). Heat treatment is desirable; titanium has been proposed for low-power machines. Steels 35 and 45, and sometimes 50 and U8, are used when the tool is integral with the concentrator.

Screwing of the concentrator to the head may set up large forces that may overstrain thin-walled supporting components, so there must be some reliable means of clamping the flange or permanent section of the concentrator for this operation. For example, one clamp may be applied to the tail of the concentrator and another to the flange.* Another method is to fix the flange with a wedge or taper system. These methods serve to protect the ways in the feed mechanism as well as the head. Some machines have provision for removing the entire vibrating system, the concentrator or tool being unscrewed with the parts held in a jig.

20. Feed Mechanisms

These mechanisms serve to apply the working force between tool and workpiece and to sustain this force during cutting.

The mechanism must have precision slides and be reasonably sensitive; precision is needed in machining to specified tolerances, while sensitivity is essential to the maintenance of the force in a specified range, which is needed to ensure the highest accuracy.

The mechanism must also include a means of reading the displacement of the head; in addition, machines for routine production use must have means of adjusting the feed force and rate in accordance with a specified program.

The earliest feed systems employed counterweights, the force being the difference between the weight of the head and that of the counterweight attached via a pulley system (Fig. 94a) or via a lever (Fig. 94b). This system has sometimes been used to move the table, in order to avoid throwing the screwing forces onto the guides [17]. The force is adjusted with variable weights or by moving the weight along the arm. Such a system is simple, but it is inconvenient to adjust and insensitive, on account of friction in the pulleys; it is also difficult to vary the force in accordance with a specified program.

The spring-loaded system (Fig. 94c) is more compact; it is also very sensitive, on account of the small number of rubbing surfaces. On the other hand, the force varies during the working movement.

Pneumatic [18] and hydraulic [19] (Fig. 94d) systems are undesirable; pneumatic methods are commonly used in high-speed machines (which ultrasonic ones are not), while hydraulic methods are mostly ones designed to supply substantial power, whereas the forces in ultrasonic work are only a few kilograms. Further, hydraulic drives do not provide good sensitivity.

Periodic switching of a motor is used in some foreign machines [20], the process being as follows (Fig. 94e).

The motor 1 is fed via contacts 2 and drives the screw 3 via a worm. The nut 4 is driven downwards by the screw and reduces the stress in the spring 5 acting on the acoustic head 6. This increases the force applied to the tool. The contacts are carried by the nut and are operated to trip the circuit when the limit stop 7 is reached. The head sinks as the cutting continues, so the contacts close again. The force may be adjusted by moving the contacts along the rod 8. The system is not reliable, because the contacts wear rapidly and soon fail.

---

* Engineer E. M. Gryaznov proposed this.

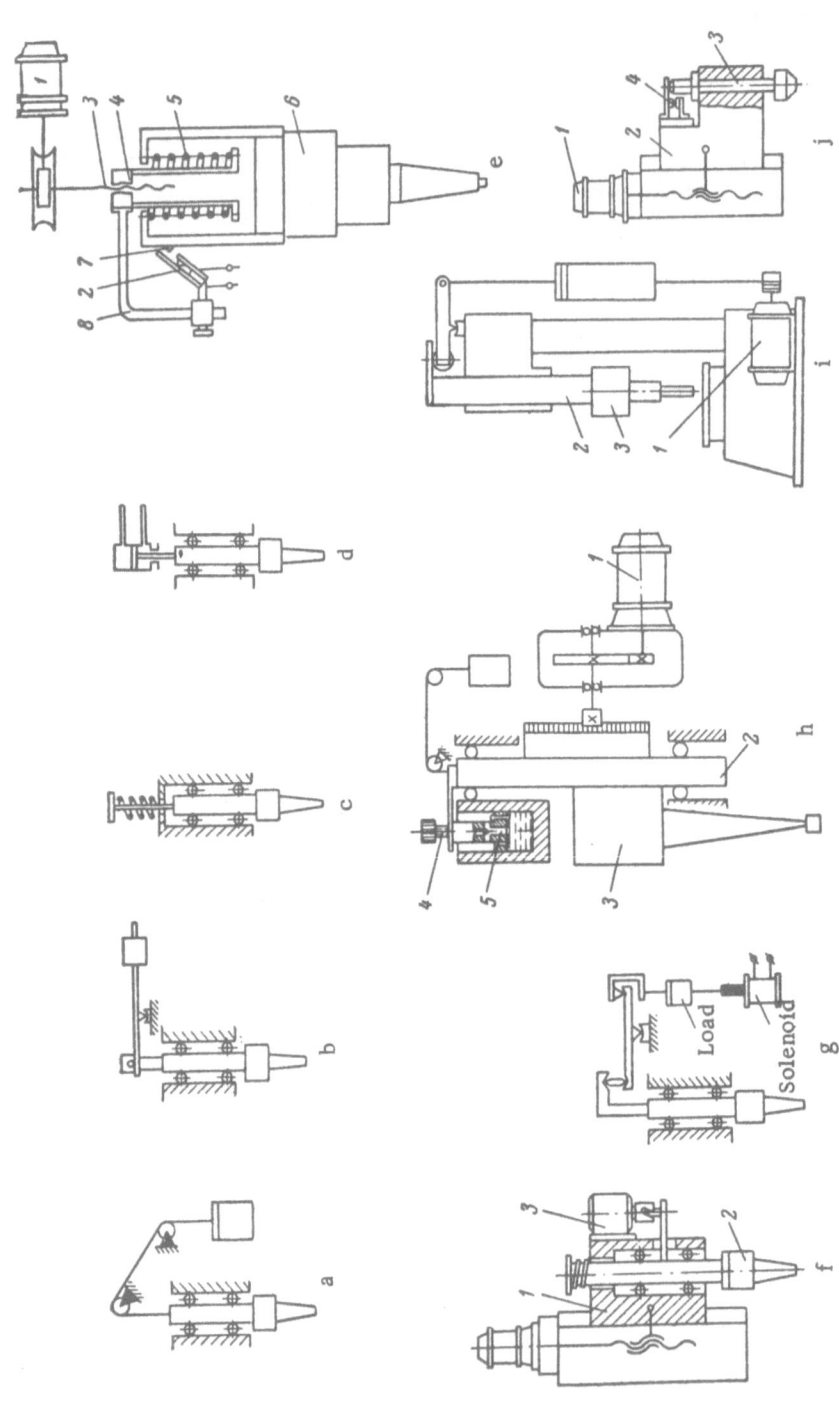

Fig. 94. Feed methods used in ultrasonic machines: a) load acting via a pulley; b) load acting via a lever; c) spring; d) pneumatic or hydraulic methods; e) motor system (with limit stop); f) inductive transducer; g) solenoid method; h) feed mechanism driven by a braked electric motor; i) the same, but with transmission via a cable; j) mechanism with forced feed.

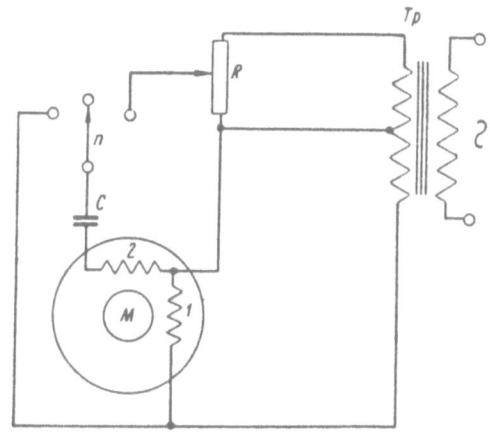

Fig. 95.   Circuit used to control feed motor.

A better system was used in the first model of the 4772, which was shown in 1958 at the Brussels Exhibition. The contacts were replaced by an inductive transducer, which recorded the displacement of the carriage 1 relative to the head 2 (Fig. 94f). Here the transducer was a selsyn, whose armature was turned through a small angle by a gear system. This method has the disadvantage of complexity [21].

The best methods, and ones particularly convenient to use, employ a solenoid or a stalled motor to control the feed mechanism.

The solenoid system (Fig. 94g) employs the force exerted by a solenoid having its core attached to the counterweight [22, 23]. The solenoid allows one to vary the force in a specified fashion by means of the current, and

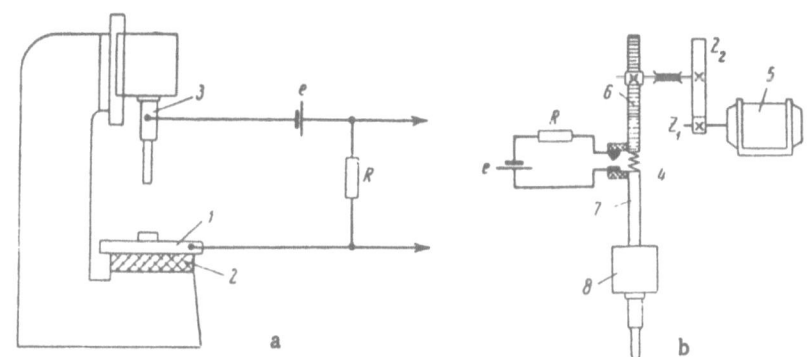

Fig. 96.   Circuits for indicating contact of tool with workpiece: a) electrical contacts; b) elastic component and kinematic system; 1) table; 2) insulation; 3) tool; 4) elastic component; 5) motor; 6) rack; 7) shaft; 8) head; e) voltage source; R) resistance.

also to perform rapid periodic lifting of the tool to allow fresh abrasive to enter the cutting zone. A slight disadvantage is that the force varies somewhat over the working stroke.

Figure 94h shows the use of a stalled motor [24]. The asynchronous servomotor 1 acts via a reduction gear and rack-and-pinion mechanism or in some other way (Fig. 94i) to move the slide 2 carrying the head 3. The stalled motor continues to develop a torque while stalled and so acts like a tensioned spiral spring.

The motor may develop too large a power if allowed to run free, so the feed is fitted with an oil dashpot, which restricts the speed of the head and makes the movement smooth.* The feed rate is adjusted with the needle valve 4, which controls the flow of oil. A relief valve 5 is fitted to produce quick return.

Figure 95 shows the motor control circuit. The stator winding 1 receives a voltage all the time, while winding 2 receives a voltage adjusted by potentiometer R. The capacitor C serves to shift the phase, while the potentiometer adjusts the torque provided by motor M and hence the force. The switch n provides feed, stop, and return actions.

This system is more complicated than the solenoid one, but it is also more flexible, and the force is independent of the distance traveled. It is also suitable for use in program control.

---

* Oil dashpots are also used in other types of feed mechanism.

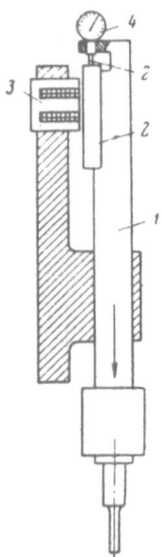

Fig. 97. Method of using
the mechanism for reading
the displacement.

Holes of diameter 0.3 mm or less make severe demands on the feed mechanism, since the working forces are 5-20 g; the frictional forces must be only 1-4 g. This is very difficult to ensure with precision ways, so there is some interest in a system providing a fixed feed rate less than the possible cutting rate. This overcomes the difficulty of sensitivity in the mechanism but measures are still needed to prevent accidental breakage of the tool. Figure 94j shows one possible method.

The motor 1 works via a worm reduction gear into the lead screw, which drives the slide 2, which bears the freely moving rod 3 carrying the tool. The rod moves up in the slide if the feed is too large, which trips the power via the contacts 4, and the motor stops.

This mechanism has been used in machines of higher power [25].

A slight tilt in the head guides has been used to provide the feed in machines of horizontal type.

The ways of the feed mechanism must be very accurate and of low friction. The accuracy is required by the high precision needed in the finished components, while low friction is needed to ensure high sensitivity. High friction means low sensitivity, which does not allow one to keep the force within the required range and involves the danger of chipping at the exit side of through holes.

Balls are often used in the guides, with two balls on each side, which are held in place by a separator. One part of the slide is adjusted by means of screws to set the working gap; best results are obtained if this part is spring loaded. Ball-bearing guides are sometimes used; these reduce the tendency to sideways movement, but the system is generally more complicated and less accurate.

All the rest of the moving system must be designed with a view to minimum friction; knife-edge systems are to be preferred to ball-bearing ones. The feed mechanism must be carefully protected from the ingress of dirt. It is usual to fit a means of reading the displacement of the tool, with various types of scale. Standard dial gages are very convenient, but these tend to raise the limit of sensitivity to 40-100 g. Their ranges are commonly 2-10 mm, whereas the working run may be 25-100 mm. A wedge system may be used, which increases the error of reading but enables one to follow the entire motion.

Induction [26] and optical systems are used if the loss of sensitivity is impermissible.

It is inconvenient to read the depth from a scale in routine production tasks, so the process must be made automatic. It is particularly important to take account of the wear of the tool, in order to ensure reproducible machining. The wear is governed by many factors, so the usual machine-tool methods are inapplicable.

The feed mechanism of an automatic or semiautomatic ultrasonic machine must perform the following cycle:

1) bring up the tool slowly to touch the workpiece without breaking the tool;
2) provide the normal cutting force;
3) reduce the force at some specified depth (in order to avoid chipping in through holes);
4) over-run a certain distance to ensure the required size of hole on the exit side;
5) return the tool to the initial position.

The lengthwise wear can be compensated if the control on the feed is actuated when the tool touches the workpiece; use can be made of the contact with a conducting workpiece or with the conducting suspension (Fig. 96a). Another method is to use the deformation of an elastic element in the feed mechanism when the tool meets the workpiece (Fig. 96b).

The system for switching on the displacement-reading system (Fig. 97) operates as follows. The slide 1 carries the magnetic stop 2; the electromagnet 3 is energized and operates the stop when the tool touches the

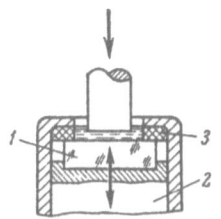

Fig. 98. Attachment by means of an elastic insert: 1) workpiece; 2) vibrating table; 3) insert.

workpiece. The shaft of the transducer 4 fixed to the slide then starts to move and records the displacement, which is used to change the feed as required [27].*

21. Tables of Ultrasonic Machine Tools

The table is used to hold the workpiece and provide coordinate displacements.

The workpiece may simply be placed on the table or may be clamped down; a magnetic chuck may sometimes be used.

T-slots and tapped holes in the table are commonly used in machine tools but are undesirable here, on account of contamination with abrasive. It is best to keep slots out of the area accessible to the suspension. The workpiece is held in an elastic insert if it must be set in vibration (Fig. 98). The loading force must exceed the inertial force developed by the workpiece. Sometimes the workpiece is soldered or cemented to the table.

Ultrasonic methods are used mainly for precision machining, so the workpiece has to be displaced by exact amounts by reference to drums or dial gages. Micrometer and optical methods are also much used.

The ways must be carefully protected from contact with the abrasive. Steep grooves or holes are made in the table to drain off the abrasive rapidly into the sump.

22. Feed Systems for the Abrasive

The suspension may be applied by hand in a low-power machine, but pumps (centrifugal or membrane type) are used in machines of higher power to supply the suspension via one or more nozzles, or via a funnel surrounding the concentrator (Fig. 99a). The latter provides not only uniform supply but also good cooling of the concentrator.

A good method is to retain the suspension in a bath in the cutting zone (Fig. 99b), which ensures a good supply and suppresses any tendency for the tool to scatter the suspension when the amplitude is large [5].

Much of the liquid supplied by the pump is used to flush the abrasive into the sump (Fig. 100). Boron carbide froths in water, so the bearings of the motor must be carefully protected.

The sump must hold 1.6-2 liters of suspension; a smaller volume may lead to failure of supply, on account of dispersal throughout the system.

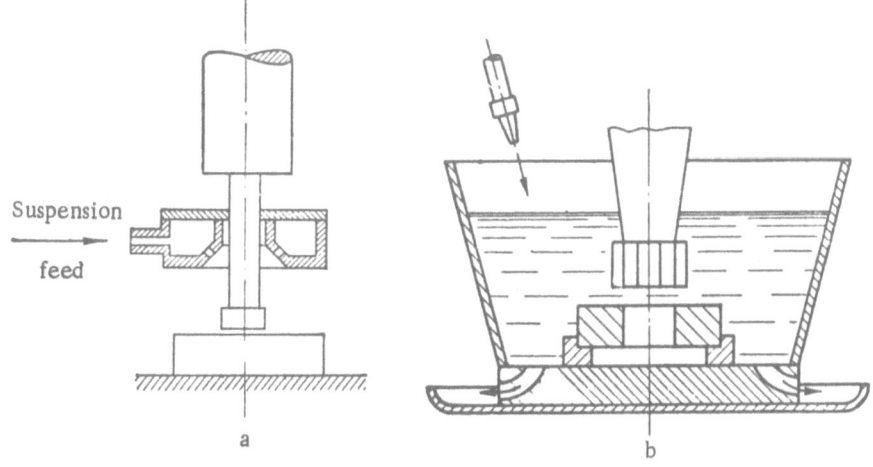

Fig. 99. Supply of suspension to the cutting zone: a) via a sprayer; b) with retention.

---

* Engineer V. E. Polotskii proposed a floating contact in the oil-dashpot system to turn on the magnet; the contact sinks and closes when the tool meets the workpiece, on account of the cessation of the flow of liquid.

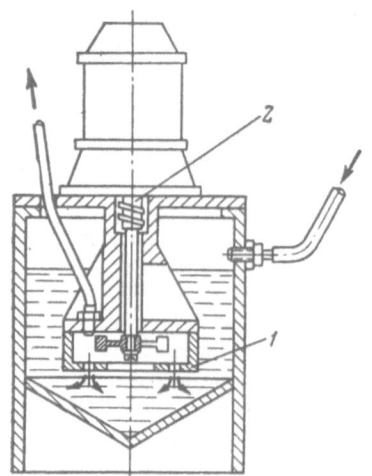

Fig. 100. Pump system
for supplying suspension.

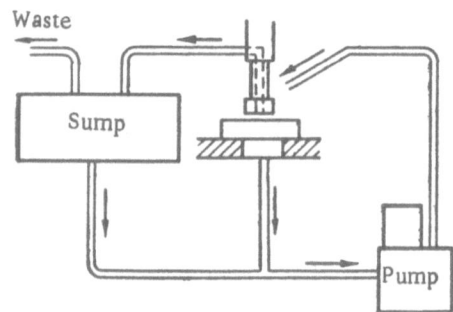

Fig. 101. System for supplying
abrasive suspension.

Some machines have provision for two successive machining operations on the same table; rough machining with coarse abrasive is done in the first, and finishing with fine abrasive in the second. Two pumps are needed in this case [28].

The suspension may be supplied or drawn off through a hollow tool or via holes in the workpiece [29-32].

In the latter case there are two pumps, one to supply the suspension to the cutting zone and the other to draw it off into an intermediate sump, which is periodically emptied (Fig. 101). Immediate recirculation can also be used.

23. Generators

Vacuum-tube generators are usually the sources for heads in ultrasonic machines; these have to satisfy the following main requirements:

1) reliability and durability;
2) efficiency;
3) simplicity in design and inexpensiveness.

Many different generators have now been made and described [33-37], so we deal only with certain features of these.

Small generators usually employ a master oscillator, a buffer amplifier, and an output stage. The disadvantages of these are the low efficiency and the numerous tubes; the advantages are the wide tuning range and the simple matching to the load. The master oscillator is often of RC type.

Transistors enable one to minimize the number of vacuum tubes and so increase the reliability. Figure 102 shows a circuit to produce 0.3-0.4 kW.

The LC oscillator and buffer amplifier employ transistors; the diodes are semiconductor ones. Only the output stage employs vacuum tubes.

The tuned circuit contains a ferrite core bearing the tuned winding (W_{IV}) and feedback windings (W_I and W_{III}), as well as the winding feeding the buffer amplifier (W_{II}).

The ferrite core provides a good Q in a small size and hence highly stable oscillations of low harmonic content. The tuning capacitor is C_1. The amplitude is adjusted with R_3.

The buffer amplifier employs P4 transistors in push-pull in the common-base connection, in order to improve the stability and minimize the effects of loading by the grid circuit of the output stage.

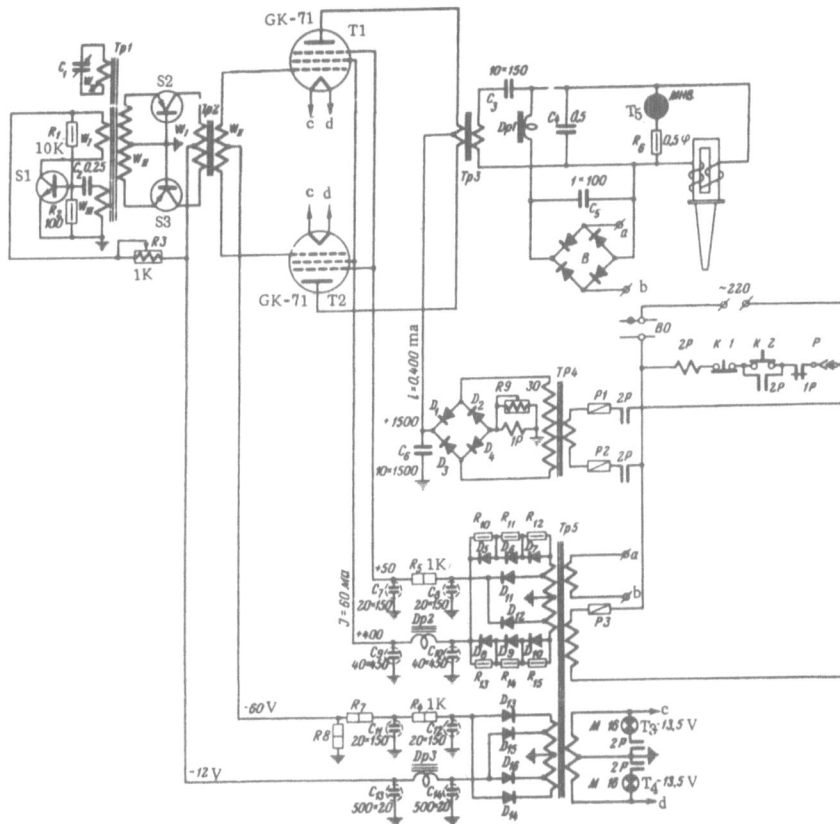

Fig. 102. Generator of output 0.4 kW, 16-25 kc model.

The output stage has two GK-71 tubes operating in push-pull, which easily provide the required power and enable one to simplify the power supply, on account of the high level of ripple permissible. The output transformer has a core of É310 steel, in order to minimize its size and losses.

Germanium diodes are used to provide the dc for the stages, but a selenium rectifier (type AVS) is used for the bias magnetization.

The rectifiers are fed from transformers Tp4 and Tp5, whose cores are also wound from É310 steel.

The signal lamps L_3, L_4, and L_5 indicate that the master oscillator is on and that power is present at the output.

Relay 1R (in the plate circuit of the output stage) operates if the plate current becomes too high; its normally closed contacts disconnect relay 2R, which disconnects the primary of the plate rectifier transformer from the line (220 V, single phase) [38].

Low-power generators have been simplified by using the dc component of the plate current for the bias magnetization [39]. This makes the generator very compact, but matching presents some difficulty. Moreover, the transducer winding is then at a high potential.

Tests have been made on the use of alternating current instead of dc for the bias magnetization; pulse generators have also been used. Neither has found practical application.

Various self-excited systems are used in high-power generators. A magnetostriction transducer represents a nonlinear load whose impedance is frequency dependent, being least at resonance (Ch. 1). Then an oscillator with inductive feedback gives a strong second harmonic, which leads to instability at the resonance frequency; jumps in the oscillator frequency virtually rule out tuning to resonance with the transducer. This difficulty is overcome by the use of current feedback, a tuned circuit being included in the feedback circuit to provide smooth frequency control. Invertor systems are also used, in which the plate circuit contains a large inductance [35].

TABLE 16. Parameters of Ultrasonic Generators Used in Machines

Parameter	Type			
	UM2-01	UM1-04	UZM-1,5S	UM1-4
Nominal output power, kW	0,1	0,4	1,5	4
Line voltage, V	220	220	220/380	220/380
No. of line phases (50 c/s)	1	1	3	3
Maximum power drain, kVA ...	0,4	1,0	2,5	8,0
Rectified plate voltage, kV ...	0,8	1,5	3,0	3,7
No. of tubes	2	2	2	2
Type of tube	GU-50	GK-74	GU-81	GU-5A
Excitation	Independent	Independent	Self-excited	Self-excited
Design frequency range	16-24	18-30	18-30	16-24
Bias current, A	3	10	10	20
Cooling of tubes	Air, natural	Air, natural	Air, forced	Water and forced air
Size, mm				
in plan	310×340	470×420	600×660	920×630
height	220	560	1450	1660
Weight, kg	25	75	275	350

Table 16 [35] gives some technical data on standard ultrasonic generators used in machines.

The resonance frequency of the vibrating system wanders on account of wear in the tool and when the tool is replaced. The frequency must be adjusted periodically in order to avoid a marked reduction in cutting rate. Various self-tuning systems have been used to eliminate the need for manual adjustment.

Positive feedback is introduced into the system (including the transducer) by means of capacitance or inductance (Fig. 103). The free end of the vibrating system bears a vibration detector, whose output voltage is fed to the input of the generator [40]. Provided that the phase and magnitude of the feedback are properly adjusted, the system responds to load changes by automatic adjustment to maximum amplitude. These self-tuning systems have come into industrial use recently, but such self-tuning does not affect the phase shift of the standing waves and so cannot compensate for increased loss in the supporting components.* Moreover, this self-tuning involves certain difficulties of design, especially for water-cooled transducers.

There are also self-tuning systems with purely circuit feedback; for instance, Fig. 104 shows a circuit with load-current feedback, the primary of the grid transformer being connected in series with the transducer winding. The maximum amplitude of vibration corresponds to resonance, and the impedance is then minimal, so the conditions are best for self-excitation when the amplitude of vibration is largest. The system therefore works stably and adjusts itself to any change in the resonance frequency of the mechanical system [42].

All such systems operate over smaller ranges of load variation than do the independently excited systems.

High frequency alternators are not now used to supply ultrasonic machine tools.

24. Industrial Models of Ultrasonic Machine Tools
Types of Machine
Universal and specialized types are distinguished, the latter being intended for some specific operation. The main parameter governing the size and power of the machine is the area to be covered, which also determines the size of the table, the distance the tool must travel, the range of feed forces, and the sensitivity of the feed mechanism.

* V. V. Ustinov and N. S. Goryachev have proposed the use of a screw connection between supporting disc and concentrator in a three half-wave system, which provides a means of adjusting the point of attachment [41]. However, it is still difficult to provide very accurate adjustment to the node.

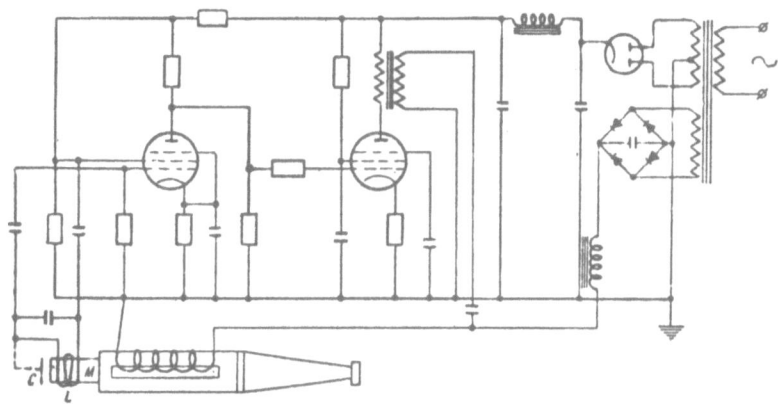

Fig. 103. Self-tuning system with inductance or capacitance coupling: M = permanent magnet, L = feedback coil, C = coupling capacitance.

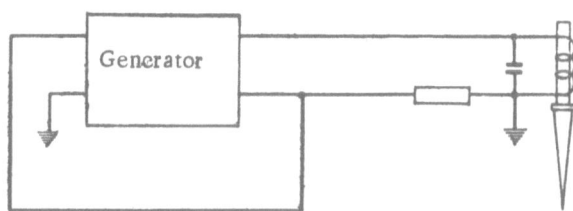

Fig. 104. System with load-current feedback.

TABLE 17. Ultrasonic Machine Tools Made in the USSR

Machine	Power, kW	Freq., kc	Hole diameter, mm	Notes
Universal 4770	0.4	18-19	0.5-10	Automatic feed control and rotating head
Universal 4772	1.5	20-33	Up to 40	With solenoid feed on head
Universal 4773	4.0	18	Up to 60	With solenoid feed on head
Universal UZS-3 (for machining hard alloys)	2.8	18-22	Up to 50	Periodic withdrawal of tool fitted
Specialized UZS-5 (for machining hard alloys)	1	18-22	Up to 50	Feed by movement of table
Experimental universal 2UPS (precision)	1-3 Interchangeable heads	16.5	—	Periodic withdrawal of tool fitted
Special 4770A	0.25	18-19	Up to 50	Semiautomatic cutter
Special MÉ-22	0.1	20	0.3-1.2	For drilling of diamond dies
Special MÉ-23	0.4	18		For cutting diamond crystals up to 2 carat
Special UZS-5M	0.5	18-20	0.8-25	For machining precious stones and minerals

The 4770, 4772, 4770A, and MÉ-22 are in regular production.

TABLE 18. Some Characteristics of Machines Made in Other Countries

Country	Firm	Model	Power, kW	Hole diameter, mm	Freq., kc	Notes
	Mullard	L.274	–	–	27	For engraving
		E. 7680/2/3	0.06	0.15-19	20	
Great Britain	Kerry	MPD	0.5	–	20	Types with two and four heads available
	Mullard and Kerry	Sonorod	2	Up to 76	20	
West Germany	Lehfeldt	Diatron A	0.5	25	20	Universal
		Diatron B				Simplified
		Diatron C				Horizontal
USA	Sheffield	Sheffield-Cavitron 100-A	0.2	12	–	Mobile, with rotating table
			–	0.076-6.35	–	With rotating tool and table feed
		Cavitron 200-B2	0.25	0.4-25.4	25	
		Cavitron 200-B2	0.6, 1.0 and 2.4 (Interchangeable heads)	0.15-90	18-80	Gravitational feed
			1	Up to 27.3	20	9-Spindle machine for use on semiconductors
USA	Raytheon	2-332	0.7	–	25	On stand
		2-333	0.3	–	–	" "
		2-334	0.1	–	–	" "
		T-2525	1.0	–	25	
France	Réalization Ultrasonik	3 Models	–	–	–	
	Sofram	–	–	–	–	
Italy	Federici	–	1.6	–	20	
East Germany	Klaman and Granert					10-Spindle machine for drilling watch jewels
Czecho-slovakia	Institute of Mechanization and Automation	VU-3	0.125	–	–	Bench type
		VU-15	–	–	–	
	Lenin works, Pilsen	U_ZV1-20	2	–	–	
China	–	–	1.5	–	22	Table 350×145 mm
Japan	Tyonpa Kogyo Co.	USM-150N	0.15	–	25 ± 10%	On stand, with hydraulic feed
		USM-500N	0.5	–	22 ± 10%	On stand, with gravitational feed
		USM-1000N	1.0	–	18 ± 10%	

TABLE 19. Rough Dimensional Series of Broaching Machines

Model	Purpose	Largest diameter, mm	Generator power, kW	Table size, mm	Max. tool travel, mm	Feed force, kg	Sensitivity, g
0	Broaching of small holes and slots in glass, ferrites, ceramics, rubies, diamonds	5-7	0.05-0.1	100×125	15	0.2	10-20 (or less)
1	Machining of small components, slicing of semiconductors	10-12	0.25-0.4	125×160	25	0.8-1.0	50-70
2	Drilling holes in ceramic components, making of small wire-drawing dies . . .	20-22	0.5-0.7	160×200	35	2.0-2.5	100-120
3	Making of small dies, machining of tools	60-80	1.0-1.5	250×350	50	Up to 6.5	±180
4	Making of wire-drawing and other dies of hard alloys, machining of hard ceramic components	100-110	2.0-4.0	400×500	50	Up to 6.5	±250

The need for model 2 has yet to be determined.

Fig. 105. The 4770 universal machine.

Tables 17 and 18 list some of the numerous machines that have been made in various parts of the world, some of them being in regular production.

The first ones were based on vertical milling or drilling machines, but the best results are obtained with specially designed machines.

The classification of machines by types is only just beginning, but the information at our disposal enables us to draw up a rough division in accordance with size (Table 19) [34].

Machines corresponding to models 1 and 3 are already in regular production (as models 4770 and 4772); a model 4 has also been designed (the 4773). Some of the most typical current machines are described below.

Universal Machines

The 4770 Universal Broaching Machine. This is a typical low-power machine mounted on a cast stand (Fig. 105), with separate generator and pump. The acoustic head employs a nickel two-rod transducer (Fig. 106). The mounting flange silver-soldered (Psr40 solder) to the end of the core has a tapped hole to take interchangeable concentrators. The flange has hexagon flats for support during changing of tools, which is performed with two wrenches, one acting on the flange and the other on the tail of the tool. The two wrenches are operated together to avoid loading the bearings and supports.

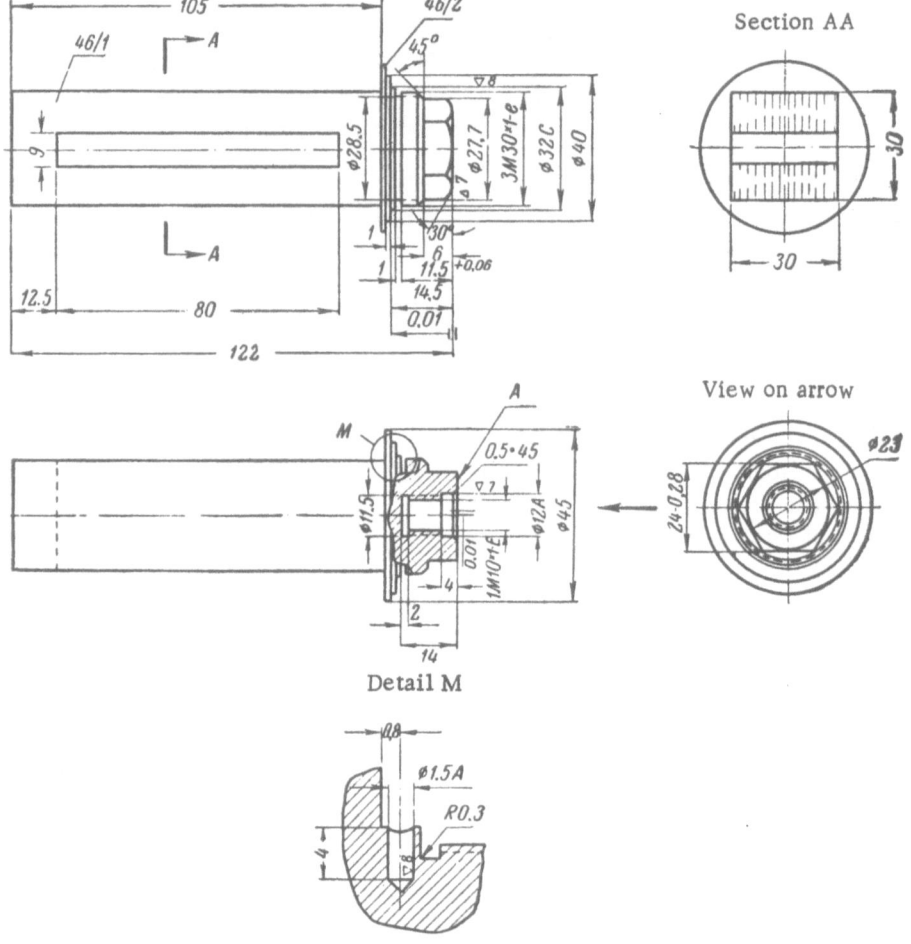

Fig. 106. The transducer of the 4770.

The magnetostriction vibrator (Fig. 107) is attached ɔ the massive carrier 6 by the thin-walled support 5 enclosing the transducer. This support is turned in the body 9 of the head by a worm drive in order to set the tool relative to the workpiece. The angular setting is read from the scale 11. The transducer is cooled by flowing water in the internal cavity, the upper end being above the water in order to minimize radiation loss. The head is attached to the slide of the feed mechanism by the clamp 13. The slide is driven by an asynchronous servomotor of ASM type via the reduction gears Z_3 and Z_4 as well as the rack and pinion 7 and 8 (Fig. 108).

The system operates as follows: the motor 2 is turned on, which displaces the slide and head downwards until the tool touches the workpiece. The stalled motor continues to develop a torque (the feed force) and ensures that the tool penetrates as cutting proceeds while maintaining the force.

The slide 1 and head are balanced by the counterweight 3. The handle 4 on the cable drum provides manual displacement. The oil dashpot restricts the rate of displacement and makes for smooth movement.

The slide may be clamped at any position; a dial gage indicates the displacement. All guides and so on employ ball bearings, to ensure high sensitivity.

The table has engraved drums to read the displacement to 0.02 mm and also engraved scales. The suspension is pumped into the cutting zone and flows away along inclined grooves and channels in and around the table. Sound design and careful manufacture of the vibrating system provide a highly reliable machine, in conjunction with the sensitive and convenient feed control; the cutting rate on glass can be 300 mm^3/min for a hole 8 mm in diameter cut with a solid tool and grade 6-10 boron carbide. The rate for T15K6 hard alloy under the same conditions is 6 mm^3/min [43-44].

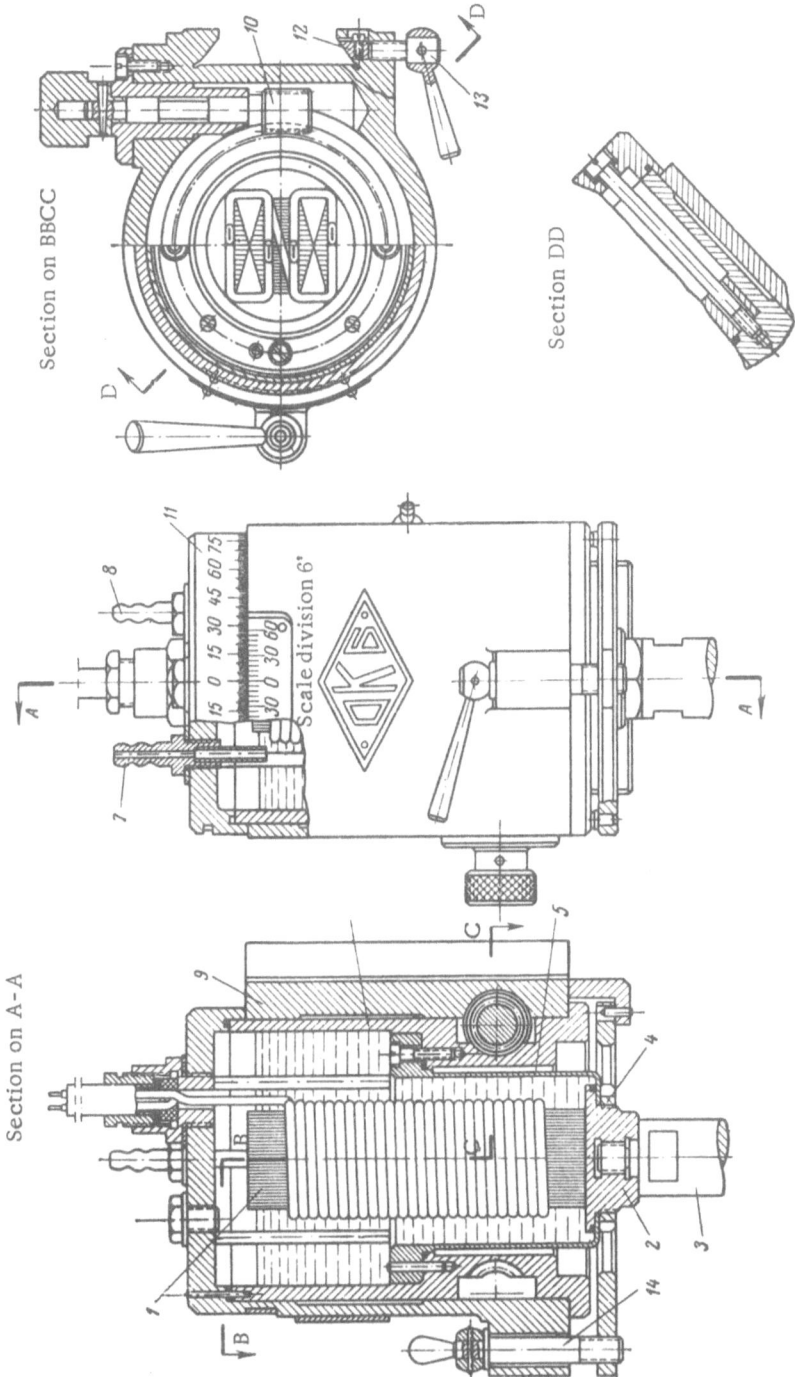

Section on BBCC

Section DD

Section on A-A

Scale division 6'

Fig. 107. The acoustic head of the 4770.

97

Diameter of holes for solid tool, mm	0.5-1.0
Maximum depth	2-5 diameters
Travel of slide, mm	100
Working range of head in slide, mm	110
Rotation of head about vertical axis, degrees	± 90
Table dimensions, mm	125×165
Maximum displacement of table, mm	
lengthwise	100
transverse	80
Feed force, kg	Up to 4.5
Sensitivity of feed, g	50-100
Frequency, kc	18-19
Range of frequency adjustment, kc	16-20
Generator power, kW	0.4
Total power drawn, kW	0.6
Dimensions, mm	515×420×705
Total weight with all parts, kg	155

The 100A Universal Machine. This is intended for use on small holes and recesses; it is a bench machine, with generator separate (Fig. 109). The head is air cooled. The vibrator is readily removed, so the tool can be set up very precisely.

The tool may be rotated at an adjustable speed and may be displaced relative to the axis of rotation, so holes and recesses of large diameter can be machined. This rotation is not used in profile machining.

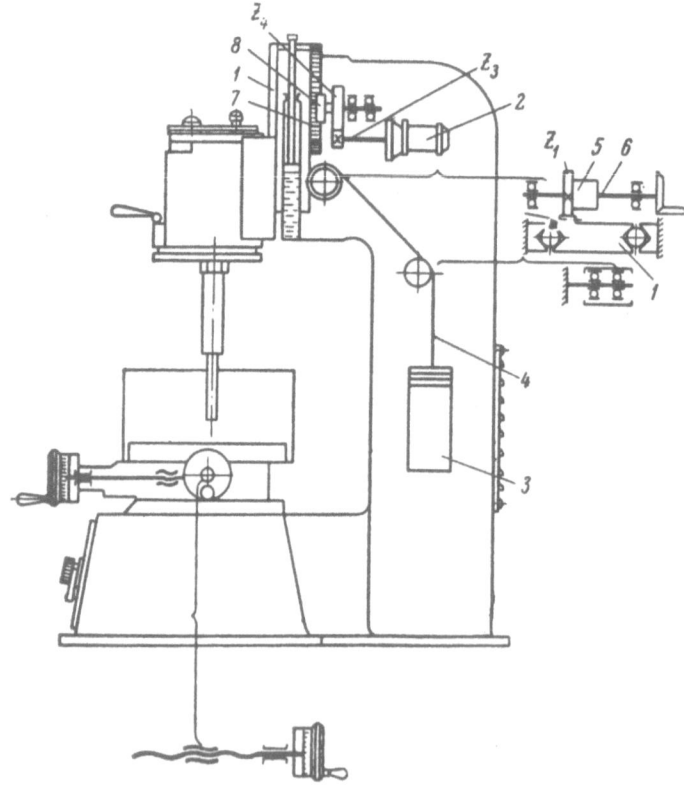

Fig. 108. Kinematic scheme of the 4770.

Fig. 109. The 100A universal ultrasonic machine.

The table has micrometer drums to read the displacements in the horizontal plane. The table also carries a chuck to hold workpieces.

The feed is provided by vertical movement of the stage in response to a spring, the vertical ways being carefully protected from contamination.

A microscope ($\times$ 30) is used to check the setting of the tool in the head; one with a magnification of 50 is used to check the setting relative to the workpiece [45].

Technical Parameters of the 100A

Diameter of holes cut, mm	0.076-6.35
Maximum diameter of workpiece, mm	69.8
Maximum height of workpiece, mm	12.7
Distance of tool from pillar, mm	69.8
Range of movement of table lengthwise and transversely, mm	15.2
Vertical feed, mm	Up to 1.27
Rotation speed of tool, rpm	200-2000
Line voltage	120 V (60 c/s)
Area required by machine, mm	Not more than 600 $\times$ 900

E. 7680/2/3 Universal Ultrasonic Machine. This is of light construction and is intended for work of medium precision (Fig. 110).

The stand consists of a cast base and vertical pillar bearing the acoustic head. Roller bearings protected by rubber sleeves are used. The displacement is read from a scale reading to 0.05 mm, which is read via a lens fitted to the cover. The moving parts are counterbalanced by a spring, whose tension is adjusted to control the feed force.

The head uses a two half-wave system with a prismatic core, the magnetic circuit being closed via two accessory armatures. The transducer is air-cooled.

The base can be fitted with a removable table not having lead screws. The top surface has holes to allow the suspension to flow into a sump.

The generator type E. 7685/2 and pump type E. 7686 form separate units. The master oscillator is of multivibrator type [46].

Fig. 110. The E.7680/2/3 universal machine. The workpiece holder is shown in the center, and the generator on the right.

Technical Parameters of the E.7680/2/3

Diameter of holes, mm 0.15-19
Depth of holes cut, mm 12.7
Table dimensions, mm .... 190×190×89
Working frequency, kc 20
Generator power, W 60
Power drain, W 175
Supplies .... 110-150 and 200-250 V
(40-100 c/s, single phase)
Pump capacity, liters 2.25
Feed force, kg .... Up to 2.5
Dimensions, mm .... 240×380×760
Weight, kg 46

The UZS-5 Ultrasonic Machine. The UZS-5 (Fig. 111) is the simplest in design and construction of all universal machines.

The transducer is made of permendur; the vibrating system is held via a flange at the node in the integral part of the concentrator (Fig. 93c).

The head has a small preset range of adjustment. The table has coordinate lead screws in the horizontal plane and can be rotated.

The vertical feed is provided by moving the table on precision ball bearings. The feed force is provided by a load, which moves the table upwards via a lever mechanism.

The suspension is supplied by a centrifugal pump in the base. The side flap takes tools, workpieces, and so on.

Fig. 111. The UZS-5 ultrasonic machine.

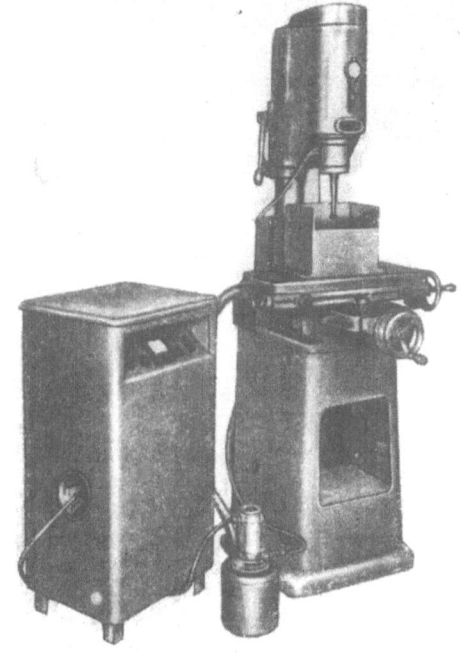

Fig. 112. The Diatron ultrasonic machine.

Technical Parameters of the UZS-5

Maximum hole diameter machined in one pass, mm . . .	50
Table diameter, mm	140
Travel of table, mm	
vertical	40
lengthwise	60
transverse	30
Angular range of table, degrees	60
Range of preset adjustment of head, mm	50
Feed force, kg	0.2-10
Sensitivity of feed mechanism, kg	0.1
Working frequency, kc	18-22
Power, kW	1-2
Dimensions of machine, mm	
length	290
width	570
height	435
Weight, kg	100

The utility of the machine is restricted by the small area of the table and the small range of the coordinate lead screws, but it has proved very valuable for making dies of hard alloys [2].

The Diatron Ultrasonic Machine. The Diatron (Fig. 112) is a medium-power universal machine. The acoustic head contains a two-rod magnetostriction transducer held by two plates attached to the core at antinodes [47]. The transducer is water-cooled. The interchangeable concentrators are attached by a locknut system.

The head is carried on ball-bearing slides; the feed is provided by an electric motor controlled as in Fig. 94e.

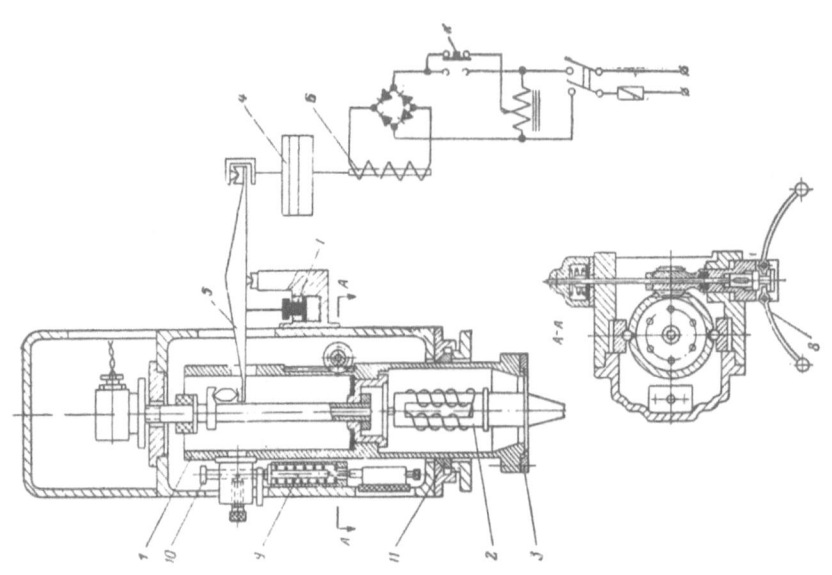

Fig. 114. The ultrasonic head and feed mechanism of the 4772.

Fig. 113. The 4772 univeral ultrasonic machine.

Later models have a pump for removing the used suspension. The generator contains only two tubes. The power is adjusted by means of the plate voltage on the output stage, a controlled rectifier with two thyratrons being used [48]. The later models have a three half-wave system attached to the body by a plate (Fig. 93c).

Technical Parameters of the Diatron

Diameter of holes, mm	0.3-5.0
Maximum depth, mm	40
Working frequency, kc	22±1
Generator power, kW	0.6
Power drain, kW	1.2
Feed force, kg	0.05-2
Pump capacity, liters	2
Vertical preset range of head, mm .	270
Mechanical head displacement, mm	35
Dimensions, mm	700×600×2220
Weight, kg	800

The 4772 Ultrasonic Machine. This is of medium power. The cast stand acts as base and link between all units (Fig. 113). The head is rigidly fixed to the upper part of the stand. Dovetail ways carry the table. The stand encloses the pump and tank for the suspension as well as the head feed mechanism and the pipes for the cooling water and suspension.

The body 1 of the head (Fig. 114) is a hollow cylinder cut away to take the magnetostriction vibrator 2, which is cooled by a water spray. The three half-wave vibrator has a nickel transducer with two rods 50×50 mm, which is held by the plate 3. The body is balanced by the counterweight 4 and arm 5 to leave an excess of 5-6.5 kg, which provides most of the feed force. The balance of the force is adjusted via the current in the solenoid 6.

This solenoid provides rapid withdrawal when the button K is pressed. Withdrawal is followed by gentle lowering, which is controlled by the oil dashpot 7, which acts in one direction only. Handwheel 8 provides manual feed. The spring 9 ensures automatic reduction of the feed force as the tool emerges from the workpiece. The rod 10 compresses this spring as the head descends and so takes the load.

Technical Parameters of the 4772

Diameter of holes, solid tool, mm	1-40
Best cutting area, mm^2	700
Maximum depth, mm	30
Travel of head, mm	50
Table size, mm	250×350
Maximum travel of table, mm	
lengthwise	100
transverse	150
vertical	150
Frequency, kc	20-22
Generator power, kW	Up to 1.5
Amplitude at end of intermediate concentrator, μ	11-11
Diameter of end of intermediate concentrator, mm	40
Length × width × height, mm	560×875×1850
Weight without generator, kg	700
Weight with generator, kg	1000

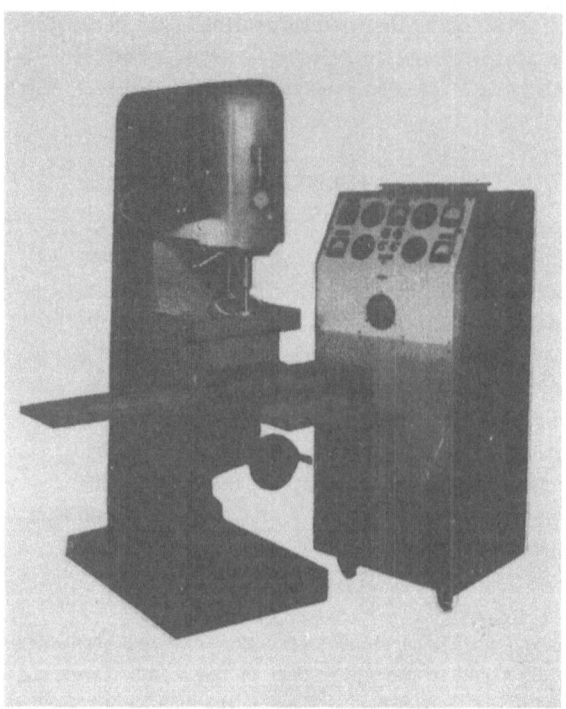

Fig. 115. The 2UPS ultrasonic machine.

This rod also carries an indicator for the degree of compression of the spring. Roller bearings allow the head to move in its support; the slides are spring-loaded. The clamp 11 links the head to the body to protect the ways from the forces exerted in screwing in the tool.

The table has 100 mm travel, which can be read to 0.01-0.02 mm. The working section of the table has numerous holes and slots, through which the suspension flows into the base of the table and thence into drainage ways and the sump. The slides are completely enclosed in jackets to keep out the abrasive.

The centrifugal pump feeds the suspension to an annular vessel with holes that encloses the tool. Most of the liquid returns along two overflow pipes to the sump; their ends contain jets that stir the suspension vigorously.

The 4772 has a high cutting rate and enables one to machine areas up to 1500 mm^2. The cutting rate on glass is up to 1200 mm^3/min for solid tools 25-30 mm in diameter used with No. 10 boron carbide. The rate for VK-20 hard alloy under analogous conditions is 20-25 mm^3/min [23].

The 2UPS Precision Ultrasonic Machine. This machine (Fig. 115) is meant mainly for making hard-alloy press tools. The acoustic head uses a four half-wave system held at two nodes in the guide transmitting the motion from the transducer to the tool (Fig. 116), which makes the head very rigid [49]. Another model has the guide supported in a plastic plain bearing (Fig. 117).

Feed at a rate below the cutting rate is used. This feed is provided by a controlled motor, which provides a fixed speed at a lever preventing the head from descending under its own weight. This type of feed reduces the chances of chipping at the exit from through holes; it is also very useful in finishing of trepanned holes.

The feed mechanism also provides for periodic withdrawal of the tool. The feed is monitored by dial gages. The slides are carefully protected.

There are two centrifugal pumps for the suspension, and provision is made for supplying the suspension via holes in the workpiece [25, 49, 50, 53].

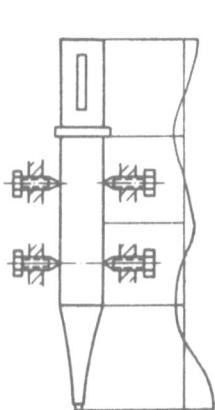

Fig. 116. Vibrating system of the 2UPS. At each node of the connecting rod there are three screws placed at intervals of 120°

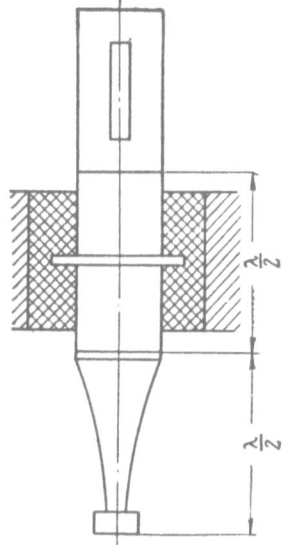

Fig. 117. Alternative form of acoustic head for the 2UPS.

Technical Parameters of the 2UPS

Table size, mm . . . . 200×400
Lengthwise travel of table, mm 300
Transverse travel, mm 150
Vertical travel, mm 150
Travel of head, mm 25
Working frequency, kc 16.5
Head power, kW 1.5
Range of feed force, kg 0-10
Sensitivity of feed mechanism, g 150
Rate of power feed, mm/min . . . . 0.1 - 2
Lifting period of head, sec 1-160
Extent of lifting of head, mm 0-25
Cutting rate on hard alloy, mm^3/min . . . . Up to 12
Dimensions, mm . . . . 1500×1300×1800
Weight, kg 400

The 4773 Universal Broaching Machine. This is of high power and is mainly intended for making hard-alloy dies and press tools.

The machine (Fig. 118) has ways for the transverse movement of the table and vertical movement of the head. The base contains two sumps with pumps to feed the suspension.

The table is moved in the two horizontal directions by handles located at the front; the positions are read from engraved scales together with verniers on the shafts (scale division 0.02 mm). Plunger pumps lubricate the lead screws and slides.

The magnetostriction transducer is mounted on a rotating plate in the head. The vibrating system is the three half-wave type. The system is held in the head by a plate fixed at a node in the stepped concentrator (carbon steel). All the moving parts work against a counterweight that leaves only 6 kg unbalanced. This residual force provides the maximum working force at the workpiece; it can be reduced by the use of a solenoid, as in the 4772.

Fig. 118. The 4773 ultrasonic machine.

Raising and lowering of the head is performed manually or automatically (by an oscillation monitor). The pillar bearing the head is driven by a built-in reduction gear and is clamped by hand in the desired position. A special jig is used to support the acoustic head when the tool is being replaced or adjusted. Six balls in a separator in the cone nut are forced in by rotation of the handle on a split ring enclosing the head. The clamping can be effected in any position.

Technical Parameters of the 4773

Diameter of holes for solid tool, mm	3-60
Best machining area, mm^2	2000
Maximum cutting depth, mm	30
Fine adjustment of head, mm	50
Coarse adjustment of ultrasonic head, mm	300
Rotation of tool, degrees	180
Table size, mm	400×500
Maximum travel of table, mm	
lengthwise	350
transverse	180
Frequency, kc	18
Power of generator, kW	Up to 4.0
Cutting rate on glass of optimal area with No. 10 boron carbide, mm^3/min	Up to 3000
The same for VK-20 hard alloy, mm^3/min . . .	30-35
Dimensions (less generator), mm	1755×1390×2350
Weight (without generator), kg	1600

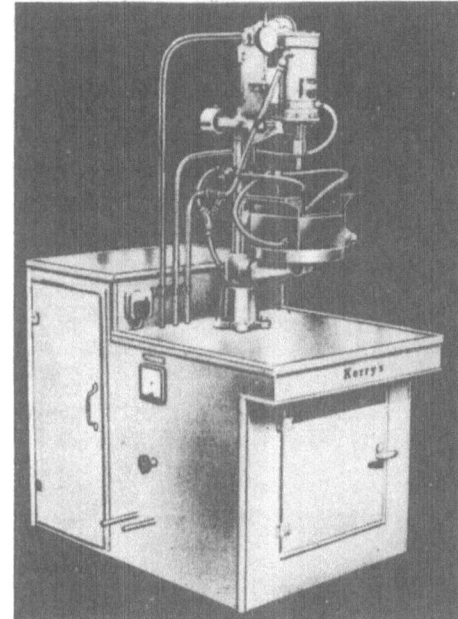

Fig. 119. The MPD ultrasonic machine: a) one-position; b) four-position.

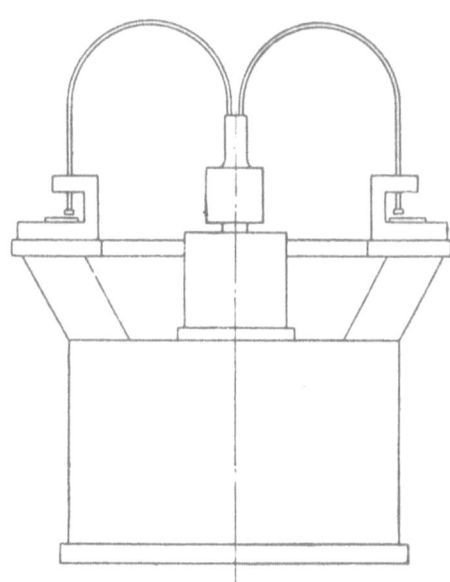

Fig. 120. System used in a multiposition machine for working semiconductors.

The oscillation monitor is located on the front face of the table. Three cams driven by a motor operate on a follower to work a microswitch, which itself controls the solenoid. The frequency of oscillation of the head (2, 4, or 6 withdrawals a minute) is set by turning a knob (this acts on the rotating cams).

The suspension feed system consists of two independent tanks (20 liters each) fitted with centrifugal pumps. These tanks are fitted with magnetic stopcocks to remove the excess water after washing of the table. These valves are set to give a working volume in the tank of 6 liters. The pump pipes near the tanks are fitted with bypass tubes to stir the suspension. Annular sprayers inject the abrasive into the cutting zone. The level in the tanks is controlled by float valves.

The flushing system is designed to remove the abrasive from the internal cavities in the lower parts of table and stand. Another branch of the water supply cools the transducer. The cooling water flow is monitored by a pressure relay interlocked with the generator.

The abrasive is flushed away via a hose that takes the water to the tank from which the suspension is supplied. The flow is controlled by two magnetic valves operated from the control panel and linked to signal lamps.

The inclined control panel lies at the front of the machine; it carries all the interlock, signal, and control gear, as well as an ammeter to monitor the force of the tool on the workpiece [50].

Technical Parameters of the MPD

Diameter of holes, mm	0.25-0.51
Accuracy of setting, mm	0.025
Machining accuracy, mm	Down to 0.012
Table size, mm	190×190×89
Frequency, kc	20
Generator power, kW	0.5
Power drawn by generator and pump, kW	2.75
Dimensions in one-position form, mm	1800×1090×788
Size of generator, mm	1066×430×635
Weight of machine, kg	450
Weight of generator, kg	206

Specialized Ultrasonic Machine Tools

Machines for Working Semiconductors. Slicing of germanium and silicon was at first done with universal machines fitted with multiblade tools, but this was not very satisfactory because such machines are rather too insensitive for use with these brittle materials. Semiconductor components are produced on a vast scale, so automatic machines are necessary.

Specialized machines for use with semiconductors have now been developed.

The 4770 has been used in the USSR as the basis for specialized machines for cutting wafers from plates of germanium and silicon. These differ from the 4770 in having a more sensitive feed mechanism, reliable protection of the feed gear from the abrasive, absence of table traverse, and semiautomatic working (tool driven slowly, increase in feed force at start of cutting, reduction of force on emergence, finishing, withdrawal). Labor productivity increased by factors of 5-8 when these machines were introduced [50]. Machines of other types and sizes have been developed for slicing semiconductors.

Specially adapted E.7680/2/3 machines are used for semiconductors in Great Britain [26]; the MPD (Kerry) may also be used (Fig. 119b).

The 2-333 is used in the USA for working semiconductors [54].

A novel development is the nine-position machine by Sheffield for semiconductors (Fig. 120). This has a single acoustic head, whose vibrations are transmitted along curved guides to the working positions, which are arranged around a circle. This machine is used with silicon wafers up to 1.27 mm thick with tools up to 27.3 mm in diameter at amplitudes of 0.025 mm at 20 kc, power 1 kW [55].

This method of transmission via long waveguides [64] appears very promising; it is found that the total cutting rate increases with the number of heads working simultaneously. For instance, the use of five tools increases the cutting rate by a factor 4. The cutting rate falls by 5-10% per half wavelength as the guides are lengthened (the more rapidly the larger the number of tools).

These tests were done on a 0.25 kW machine at 25 kc. The combined waveguide and tool in each position was a steel wire 1 mm in diameter and up to 1 m long. Holes were drilled in plates of glass and steatite with a water suspension of 220 mesh boron carbide [56].

Fig. 121. Ultrasonic machine for drilling watch jewels.

Machine for Drilling Watch Jewels. Klamen and Granert (Dresden) make the ten-position machine shown in Fig. 121 for drilling small holes in watch jewels.

In this case the blanks are set in motion by attachment to the ends of transducers. The tools are pieces of wire soft-soldered into holders, whose weight provides the feed force. These holders have electric heaters to provide rapid wire feed and soldering. The tool does not rotate. The hourly output of such a machine is 1000 stones, though this is still below the output of machines for mechanical drilling [57, 58].

Ultrasonic Dental Equipment and Engraving. Apparatus. An ultrasonic head for hand use can be employed in engraving and in the working of hard and brittle materials, as well as in dentistry [49, 59-61].

The apparatus (Fig. 122) consists of a manual acoustic head* and a generator. The magnetostriction part is the nickel half-wave tube 1 (Fig. 123) slit along its length to minimize heating. The half-wave concentrator 2 (brass) is soldered to the tube and has the interchangeable tool 3 screwed to its other end. The vibrating system is held at two nodes. The bias and exciting windings are mounted on a brass cylinder, also slit to minimize heating. The coil consists of PÉV 0.2-0.3 mm wire, three layers. The former and coil are resin-impregnated.

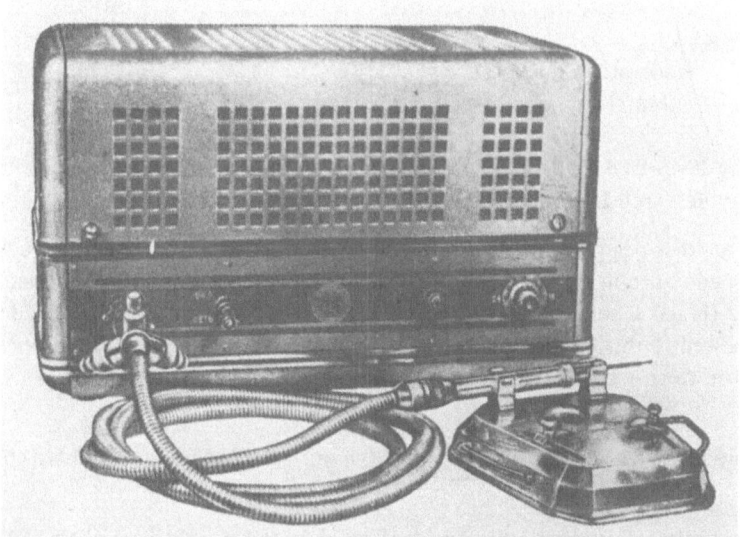

Fig. 122. Apparatus for ultrasonic engraving.

---

* One such head is 178 mm long and 19 mm in diameter, and weighs 283 g [61].

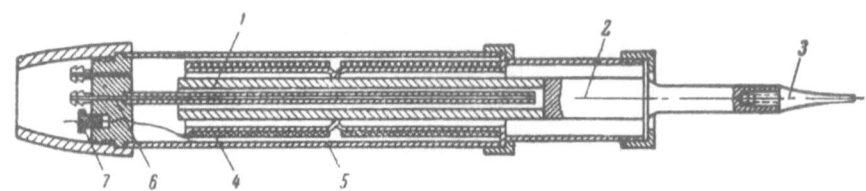

Fig. 123. The acoustic head of the engraving apparatus.

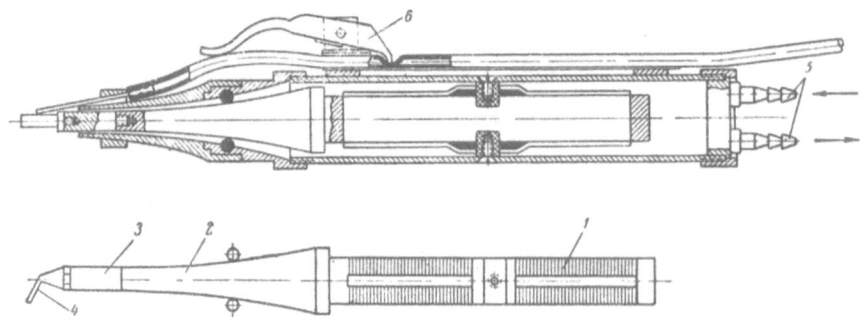

Fig. 124. Acoustic head for dentistry.

The outer cover 5 is made of mild steel to minimize leakage of the magnetic flux. The vibrator is cooled by running water, which enters along pipe 6 and passes through the vibrator. The cable 7 and the pipes for the water are enclosed in a flexible metal sheath.

The generator type UG-32 has a power of 200 W. The supply to the head is controlled by a pedal.

Technical Parameters of the Ultrasonic Engraving Apparatus

Resonance frequency, kc	26
Frequency range of generator, kc	23-29
Dimensions, mm	
generator	450×262×310
vibrator	21×195
Weight, kg	
generator	28
vibrator	0.25

This apparatus was developed by the Acoustics Institute, Academy of Sciences of the USSR, in collaboration with the Technology Research Institute, Leningrad.

Figure 124 shows another design of manual head; here a two half-wave system is used, which is held at the nodes in transducer and concentrator. The two-rod nickel transducer 1 is cooled by running water. The exponential concentrator 2 (brass) is soft-soldered to the core. The tip 3 is made of hardened steel; the interchangeable tool 4 is screwed to this. The pipes 5 carry the water flow. There is also a pipe bringing in the abrasive suspension, whose flow is controlled by the sprung lever 6. The suspension flows when the lever is depressed.

The working frequency is 25 kc; the amplitude is 20 μ for an input power of 40 W [40, 62].

Literature Cited

1. A. I. Markov, "Kinematics of ultrasonic dimensional machining," Stanki i Instr. No. 10:15, 1959.
2. L. D. Rozenberg and D. F. Yakhimovich, Ultrasonic Dimensional Machining of Brittle Materials, Profizdat, 1961.

3. "Ultraschallgeräte- und Anlagen zum Prüfen, Reinigen, Bohren, Schweissen, sowie für die Brauerei zum Extrahieren von Hopfen und im Abfüllautomat zum Entlüften der Bierflaschen," Leaflet of Dr. Lehfeldt und Co.

4. N. V. Klimushinskii, Method of cutting hard and brittle materials, Author's certificate, USSR No. 115439, dated February 5, 1958.

5. N. S. Goryachev, Production of Hard-Alloy Dies by Ultrasonic Machining, Izd. Mosk. doma nauchno-tekhnicheskoi propagandy im. F. É. Dzerzhinskogo, 1957.

6. J. H. Greening, High frequency honing, US Patent No. 2939251, dated June 7, 1960.

7. P. E. D'yachenko and Yu. N. Mizrokhi, Method of cutting diamonds, Authors' certificate, USSR No. 123866, dated March 19, 1959.

8. V. N. Barke, Method of ultrasonic slicing, Author's certificate, USSR No. 121644, dated Dec. 29, 1956.

9. Fukumoto, Machinery (Japan) 22(8):1281, 1959.

10. M. N. Gumanyuk, "Use of ultrasonics in technological processes," Byul. Tekhn.-Ékon. Inform. Sov. Nar. Khoz. Khar'kovskogo Ékon. Admin. Raiona, No. 1:26, 1958.

11. New machine for ultrasonic drilling, Abstract, Tekhn.-Inform. Byul., Moscow, TsINTIMash, No. 8:4, 1959.

12. L. A. Petermann, Ultrasonic drill, US Patent No. 2834158, dated May 13, 1958.

13. D. F. Yakhimovich, "Construction and design of the vibrating systems of acoustic heads for ultrasonic machine tools," in: Advanced Scientific and Production Experience, No. M-59-418/10, Izd. TsITÉIN, 1959.

14. G. I. Glazov and D. F. Yakhimovich, "Acoustic head for an ultrasonic machine tool," in: Advanced Scientific and Production Experience, No. M-59-418/10, Izd. TsITÉIN, 1959.

15. C. F. Brockelsby, "High power transducers for frequencies around 20 kc/s," Commun. congr. internat. sur les traitements par les ultrasons, Marseille, 1955, p. 113.

16. V. V. Ustinov, "Manufacture of magnetostriction transducers for ultrasonic machine tools for working hard alloys," in: Advances in Electrical and Ultrasonic Machining of Materials, Lenizdat, 1959, p. 195.

17. L. Balamuth, Method and means for removing material from a solid body, US Patent No. 2580716, dated Jan. 11, 1951.

18. E. A. Neppiras, "Report on ultrasonic machining," Metalwork. Product. 100(27-31):33, 34, 1956.

19. V. Zaguzov, "Laboratory ultrasonic apparatus," Prom. Altai. Sov. Nar. Khoz. Altaiskogo Ékon. Admin. Raiona, No. 4:28, 1958.

20. G. C. Brown and R. N. Roney, Machine device, US Patent No. 2942383, dated Oct. 17, 1957.

21. V. N. Barke and A. L. Livshits, "Current situation and trends in the ultrasonic machining of materials," in: Current Trends in Machine-Tool Technology, Mashgiz, 1957, p. 152.

22. I. V. Stroganov, N. A. Kudrin, and N. V. Trofimov, Machine for polishing and forming holes by erosion in components consisting of hard minerals or alloys, Authors' certificate, USSR No. 114937, dated Oct. 31, 1957.

23. A. L. Livshits, B. Kh. Mechetner and V. N. Barke, "The 4772 universal ultrasonic machine tool," Stanki i Instr. No. 6:10, 1959.

24. N. I. Blitshtein and D. F. Yakhimovich, Ultrasonic machine tool for working hard and brittle materials, Authors' certificate, USSR No. 117882, dated June 10, 1957.

25. V. Yu. Veroman, "Ultrasonic method of making hard-alloy dies," Advanced Scientific and Production Experience, No. M-60-29/2, Izd. TsITÉIN, 1960.

26. R. D. Knight, "Production of islands and dice in semi-conductor slices with an ultrasonic drill," J. Sci. Instr. 37:263, 1960.

27. E. M. Gryaznov, A. S. Zhivitskii, and D. F. Yakhimovich, Ultrasonic machine tool, Authors' certificate, USSR No. 129932, dated Oct. 27, 1959.

28. M. A. Raznitsyn, Machine for ultrasonic working, Author's certificate, USSR No. 122955, dated March 21, 1959.

29. C. J. Thatcher and B. Carlin, Tools for ultrasonic cutting, US patent No. 2774193, dated Oct. 10, 1955.

30. C. J. Thatcher, Ultrasonic tools, US patent No. 2774194, dated Nov. 11, 1954.

31. "Ultraschall-Bohrmaschine Diatron," Leaflet of Dr. Lehfeldt und Co.

32. C. J. Thatcher, High speed machining by ultrasonic impact abrasion, US patent No. 2804724, dated Sept. 3, 1957.

33. D. A. Gershgal and V. M. Fridman, Ultrasonic Equipment, Moscow-Leningrad, Gosenergoizdat, 1961.

34. L. D. Rozenberg and D. F. Yakhimovich, "Current position and development prospects in ultrasonic cutting, in: Current Situation and Trends in Machine-Tool and Instrument Technology, Mashgiz, 1960, p. 260.

35. D. B. Mondrus and I. M. Solomakhin, "Ultrasonic generators designed by TsKB UVU," in: Sources of Ultrasonic Energy, Moscow, Izd. TsINTI Elektrotekhn. Prom. i Priborostr., 1960, p. 5.

36. N. A. Belousov, V. P. Volodin, M. M. Zaretskii, and E. M. Shlenskii, "Aspects of the circuits and design of industrial ultrasonic generators," in: Sources of Ultrasonic Energy, Moscow, Izd. TsINTI Elektrotekhn. Prom. i Priborostr., 1960, p. 18.

37. Yu. I. Kitaigorodskii and M. G. Kogan, A single-stage self-excited ultrasonic generator, Authors' certificate, USSR No. 115906, dated March 28, 1957.

38. N. I. Blitshtein, "Circuit of a 400 W ultrasonic generator," in: Sources of Ultrasonic Energy, Moscow, Izd. TsINTI Elektrotekhn. Prom. i Priborostr., 1960, p. 52.

39. Dispositif d'excitation et de polarisation de vibrateurs a magnétostriction, French patent No. 1118506, dated June 14, 1957.

40. M. V. Ardenne and H. Rackwitz," Über eine Ultraschall-Zahnbohreinrichtung," Nachrichtentechnik 8(10), 1958.

41. V. V. Ustinov and N. S. Goryachev, A velocity transformer, Authors' certificate, USSR No. 105398, dated March 29, 1956.

42. A. M. Borok and G. S. Kratysh, "Some circuits of ultrasonic vacuum-tube generators," in: Sources of Ultrasonic Energy, Moscow, Izd. TsINTI Elektrotekhn. Prom. i Priborostr., 1960, p. 28.

43. N. I. Blitshtein, G. I. Glazov, and D. F. Yakhimovich, "The new ultrasonic machine tool model 4770," Akust. Zh. 5(1):117, 1959.

44. D. F. Yakhimovich, "The 4770 universal broaching machine," Stanki i Instr. No. 6:11, 1961.

45. "Ultra-precise ultrasonic machine tool," The Sheffielder, No. 3, 1959.

46. Ultrasonic drill and generator, Brochure by Mullard, Britain.

47. Ultrasonic vibratory device, Canadian patent No. 510791, dated May 18, 1950.

48. "L'usinage des métaux et corps durs sur la machine ultra-sonore Diatron," Ind. franc. Achats et entret. matér. industr. 7(78):891, 895, 897, 899, 1958.

49. V. Yu. Veroman, "Attachment of vibratory systems in the heads of ultrasonic machine tools," Stanki i Instr. No. 2:13, 1960.

50. B. Kh. Mechetner, A. A. Ust'yantsev, and D. F. Yakhimovich, "Universal and specialized ultrasonic machine tools," in: Use of Ultrasonics in Machine-Tool Technology, Izd. TsINTI Elektrotekhn. Prom. i Priborostr., 1960, p. 197.

51. V. Yu. Veroman, Dimensional Ultrasonic Machining of Materials, Mashgiz, 1961.

52. V. Yu. Veroman, "Attachment of vibratory systems in the heads of ultrasonic machine tools," Stanki i Instr. No. 2:13, 1960.

53. "Ultrasonic machining equipment types MPD1, 2 and 3," Kerry's (Ultrasonics) Ltd., Britain.

54. "Machines d'usinage rapide par ultra-sons Raytheon," Machine Moderne 55(626):105, 1961.

55. "Ultrasonics will go round corners," Metalwork. Product. 104(24):169, 1960.

56. G. Nishimura and S. Shimakawa, "Ultrasonic mechanical machining. III. Multi-tool ultrasonic mechanical machining, J. Fac. Eng., Univ. Tokyo 25(1):51, 1947.

57. J. Matauschek, Einführung in die Ultraschalltechnik, Berlin, 1957, p. 432.

58. E. Jakob, "Die Herstellung von Uhrensteinen," Monatsschr. Feinmechanik u. Optik 76(1):18, 1959.

59. A. A. Kolosov and A. R. Livenson, "Use of ultrasonics in stomatology," Med. Promyshl. SSSR, No. 5, 1959.

60. L. Balamuth, "Technical aspects of the Cavitron process in dentistry," IRE Conven. Rec. March 21-24, pp. 9, 89, 1955.

61. E. A. Neppiras, Design of Ultrasonic Machine Tools, Ed. by Inst. Mechan. Eng., London, 1958.

62. M. v. Ardenne, H. Grossmann, H. Rackwitz, J. Matauschek, and W. Steglich, Über einige Forschritte bei der Entwicklung von Ultraschall-Zahnbohrgeräten," Dtsch. Stomatol. 10(2):127, 1960.

63. R. N. Roney, Machine tool, US patent No. 3015914, dated June 19, 1959.

64. R. N. Roney, Ultrasonic machine, US patent No. 3027690, dated Nov. 20, 1959.

CHAPTER 5

TECHNOLOGY OF ULTRASONIC MACHINING

25. Technological Parameters*

Workable Materials

The following brittle materials can be worked efficiently by ultrasonic methods:

Agate	Glass†	Nephrite
Alabaster	Glass-micanite	Onyx
Barium titanate	Granite	Porcelain
Boron carbide	Graphite	Quartz (crystalline
(sintered)	Gypsum	and fused)†
Ceramics†	Hard alloys†	Rock crystal
Corundum	(tungsten and	Ruby
Diamond	titanium carbides)	Sapphire
Earthenware	Jadeite	Silicon†
Ferrites†	Jasper	Steatite
Fluorite	Marble	Thermocorundum
Germanium†	Mother of pearl	Tourmaline
		Zirconium boride

†Denotes a material on which there are particularly many papers, on account of industrial requirements.

A good finish and high precision require that the material of the workpiece should not dissolve in the suspension or react with it. Electrical conductivity is unimportant.

In the first few years after the introduction of ultrasonic methods there were many reports of their use on heat-resisting, hardened, tool, and (even) stainless steels, as well as magnetic alloys of alnico type, tungsten, molybdenum, and so on [5].

However, the low cutting rates and rapid tool wear resulted in no widespread application of these methods to such materials. It was found more economical to use electrical erosion methods.

In principle, all materials could be worked by ultrasonic methods at low temperatures, when they become brittle, but this involves great technical difficulties [3, 6].

Tools

The tool wears as a result of contact with the abrasive, which tends to erode the tool; cavitation and other such effects also affect the tool. Most of the wear occurs at the end, with wear at the sides being some ten times less. It is usual to specify the wear as a percentage of the depth cut. The wear is proportional to the working time [7]; materials differ in their cutting rates, so the tool wear is governed by the workpiece material, other things being equal (Table 20).

Brass, duralumin, and tungsten are unsuitable for the tool on account of extremely great wear; copper is unsuitable on account of plastic deformation [8]. Steel is the usual material, though views differ on the grade of steel and method of heat treatment. Some evidence indicates that soft low-carbon steel 20 is the best [9],

*See [1-4].

TABLE 20. Linear Wear of a Steel Tool on Various Materials [1, 12]

Material	Wear as % of cut	Material	Wear as % of cut
Germanium	0.5	Quartz	2.0
Silicon	0.5	Agate	2.9
Ferrites	0.5	Corundum	12
Ceramic (not hard)	0.7	Ruby	50
Optical glass	0.9	Tungsten carbide hard alloys ..	60-80
Soda glass	1.0	Hardened steel	100

but other results indicate that the harder tool steels and even hard alloys are better (Table 21), with heat treatment being unimportant [10, 11]. *

These differences arise from lack of knowledge of the process, which is governed by many factors. In general, we may say that the parameters of the steel are not critical and that the grade of steel can be chosen from purely practical considerations.

The tool wear on hard alloy is so great that each tool is good only for about one hole; it is best to use cheap types of steel in such cases. Low-carbon untreated steels are suitable, such as types 20 and 35. However, it is essential to use a steel of higher carbon content when the tool is integral with the acoustic transformer, since these have lower frictional losses (e.g., grades 40, 40Kh, 45, 50, and 65G). Tools for drilling small holes are made from steels 70 and U8A.

Stainless steels are highly resistant to cavitation erosion but it is difficult to make tools of such steels on account of machining troubles.

Heat-treatment of the tools is seldom used nowadays, for it complicates production, especially for profile tools.

The wear is not great if the workpiece is of glass, semiconductor, and so on, and if the abrasive (corundum) is not of high strength. In this case the tool can be made of hard alloy, which is highly resistant to wear [3, 14]. On the other hand, such tools are somewhat troublesome to make.

Allowance must be made for the effects of tool size and mass on the resonance frequency and concentrator size; the frequency falls as the mass and length increase, and conversely.

Chapter 3 deals with the theoretical side of this; here we consider some purely practical aspects.

TABLE 21. Wear on Tools of Various Materials when Cutting Glass [10]. Abrasive boron carbide No. 10; pressure 0.15 kg/mm^2; outside diameter of tool 7 mm; inside diameter of tool 4 mm; depth of hole 15 mm; amplitude 0.02 mm; frequency 18 kc; workpiece material ceramic)

Tool material	Rel. length wear of tool, %	Max. diam. wear of tool, mm	Cutting rate, mm/min
VK8 hard alloy	1.3	0.03	1.5
R9 steel	1.7	0.068	0.7
U8 steel not heat-treated	2.5	0.082	0.6
U8 steel heat-treated	2.8	0.046	1.4
Steel 45	2.7	0.099	0.6
Kh18N9 steel	3.5	0.089	0.44
Brass	15	–	–
Aluminum	30	–	–

$R_A = 90$, $\sigma_c = 70$ kg/mm^2, $\sigma_b = 10-20$ kg/mm^2.

---

* The cutting rate increases somewhat with the hardness of the tool (Table 21), probably on account of greater force in the blows.

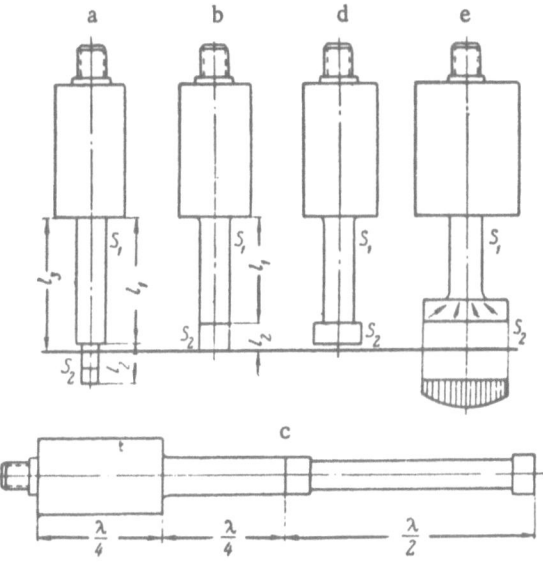

Fig. 125. Examples of connections between
concentrator and tool.

There are five possible cases [11]:

1) the tool alters the length of the concentrator, but its mass is small and its cross section is less than that of the concentrator (Fig. 125a);

2) the tool alters the length of the concentrator and the two cross sections are comparable at the junction (Fig. 125b);

3) the length of the tool is about a half wavelength (Fig. 125c);

4) the tool is a localized mass whose transverse dimensions are much less than the wavelength (Fig. 125d); and

5) the transverse dimensions are comparable with the wavelength (Fig. 125e).

The length of the tool may be neglected in the first case; but this case approaches the second one as the diameter increases, so the physical length must be replaced by the equivalent length in the calculation (Fig. 125a):

$$l_e = l_1 + l_2 \, S_2/S_1. \tag{5.1}$$

In the second case the length of the tool is included in that of the concentrator.

In the third case it is best to make the tool resonant (length a multiple of $\lambda/2$). Such tools are suitable for drilling deep holes, but their strength and precision are poor.

In the fourth case the concentrator should be shortened, the mass of the removed part being equal to that of the tool. This gives satisfactory results if the tool is not longer than $\lambda/10$ (mass not more than $\rho S\lambda/10 = \rho cS/10f$, in which ρ, S, and λ refer to the concentrator, ρ being the density) [14].

The fifth case is the most complicated and can scarcely be discussed analytically. The amplitude may not be the same at all points on the end of the tool, and transverse vibrations may occur, which may result in a marked fall in cutting rate (including complete cessation, if $\xi = 0$ at any point), oversize holes, and fracture of tool.

It is generally desirable for the diameter of the circle circumscribed about the tool to be not more than 1.5-2 times the diameter of the end of the concentrator.

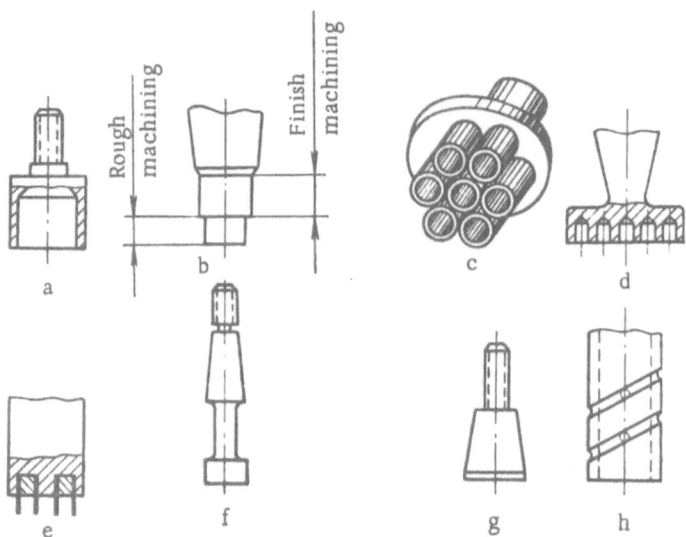

Fig. 126. Tools used in ultrasonic machining: a) hollow; b) stepped;
c) assembled from tubes; d) with holes drilled in end; e) carrying blades;
f) slotted; g) recesses behind the head; h) with flutes and holes for the
suspension.

Transverse vibrations produce oversize holes and uneven wear on the tool; they can be suppressed for complex profile tools by locating the center of gravity on the axis of the vibrating system.* The tool should also be as short and rigid as possible.

The tool is made hollow, if possible, in order to reduce the volume to be cut in broaching (Fig. 126a). This increases the linear cutting rate, since not all the material needs to be broken up; moreover, the trepanned material may be useful.

The internal contour should be parallel to the external one if possible, in order to ensure even wear. Electroerosion may be used to form holes of special shape in the tool.

On the other hand, excessive lightening of the tool by removal of the center may lead to poor rigidity and may cause an uneven distribution of the amplitude around the circumference.

A thin-walled tool should not have deep scratches on its surfaces remaining from its own production, since these may cause fracture. The outer and inner surfaces should be finished by grinding or polishing.

The thickness of any wall or projection should be at least 5 times the grain size of the abrasive; if it is not, such parts may become sharpened and wear away [13]:

Width of cutting part, mm	0.25	1.0	1.5	Over 2
Wear, %	200	70-80	60	50

The walls of a hollow tool should not be thinner than 0.5-0.8 mm, except when workpieces of costly materials are used, when high wastage is impermissible.

The working life of the tool may be extended by making it somewhat longer than the minimum necessary; this will lower the resonance frequency below the nominal one for the machine. The frequency gradually increases as the tool wears, so the amplitude at first rises and then falls again. The tool is used until the amplitude has fallen too far or the wear has begun to affect the accuracy.

---

* Guides are sometimes used to suppress transverse vibrations.

Deviations of ±6% from the nominal frequency (e.g., 1200 c/s at 20 kc) are permissible for machines of medium power (1.5 kW). This corresponds to a change in length of the tool by 12% of a wavelength [15]:

$$\Delta L = \frac{\lambda}{8} = \frac{c}{8f} . \tag{5.2}$$

The permissible deviation is smaller for low-power machines (0.1-0.4 kW), so care must be taken in tool design. Any new tool should be designed with an allowance of 3-5 mm for frequency adjustment. This is machined away if the resonance frequency is too low when the tool is fitted to the head.

If the resonance frequency is too high when the new tool is fitted (concentrator too short), it can be lowered to the required value by reducing the cross section of the lower part of the concentrator.

Tool sizes should be calculated with allowance for the oversize holes that are produced (side clearance 0.06-0.36 mm, dependent on grain size of abrasive; see below).

If a hole is to be machined successively by several tools in order to reduce the tolerance, each tool should leave a margin to be taken up by the next.

Holes in thin workpieces are best cut with stepped tools (Fig. 126b), as this eliminates the operation of tool changing.

Multiblade tools are very useful; for example, small round blanks may be cut with a tool assembled from a large number of small tubes (Fig. 126c) or with a tool having holes drilled in the end (Fig. 126d). Slicing into plates may be done with a tool having a set of parallel blades (Fig. 126e). These tools are not particularly easy to make.

The tool design must allow the abrasive to enter the cutting zone as freely as possible. Flutes, slots, and so on (Fig. 126, f-h) are used to ensure this.

The tool must be accurately made, in accordance with the error allowed in the operation. For example, a permissible error of 0.01-0.02 mm requires a tool of tolerance 0.005-0.008 mm.

The surface finish of such a tool should not be below class 7 or 8.

Worn tools are discarded; for example, a tool used on hard alloy may serve for 1-10 workpieces, whereas one used in engraving glass may serve for 300. Interchangeable tools are essential if large numbers of a particular type of workpiece are to be used.

The tool is usually screwed to the concentrator, with a cylindrical or taper neck for centering (Fig. 127a and b). The screwed part should not be less than 5 mm in diameter, since fracture becomes too probable. Soft-soldered joints give the best results, but there may then be some difficulty in setting the tool relative to the axis of the concentrator, which can be inconvenient if the operation involves the sequential use of several tools.* A shoulder and clamping ring is more convenient (Fig. 127c and d), as is a drawbolt system; the latter is better than the former, which tends to restrict the tool diameter and whose thread tends to become contaminated with abrasive. On the other hand, a drawbolt may involve removal of the concentrator or a special hole through the head, which is not always possible.

A simple taper shank (Fig. 127e) or soft soldering may be sufficient if the power is low.

Tools up to 3-5 mm in diameter may be held in a chuck, though this tends to increase the frictional losses. The needles used in drilling diamond dies are held in this way, since they must be replaced frequently. Figure 127f-h illustrates such methods.

The tool may be hard-soldered to the concentrator (PSr45 or PSr40KN solder), which gives a very reliable joint but which is inconvenient, on account of the soldering operation. The required accuracy of tool setting may involve the use of a jig and even of final machining of the tool after attachment. The most reliable but least convenient tool is one integral with the concentrator.

---

* Sometimes the tool has two locating pins or other such components, which are used to set the workpiece relative to the tool after the change.

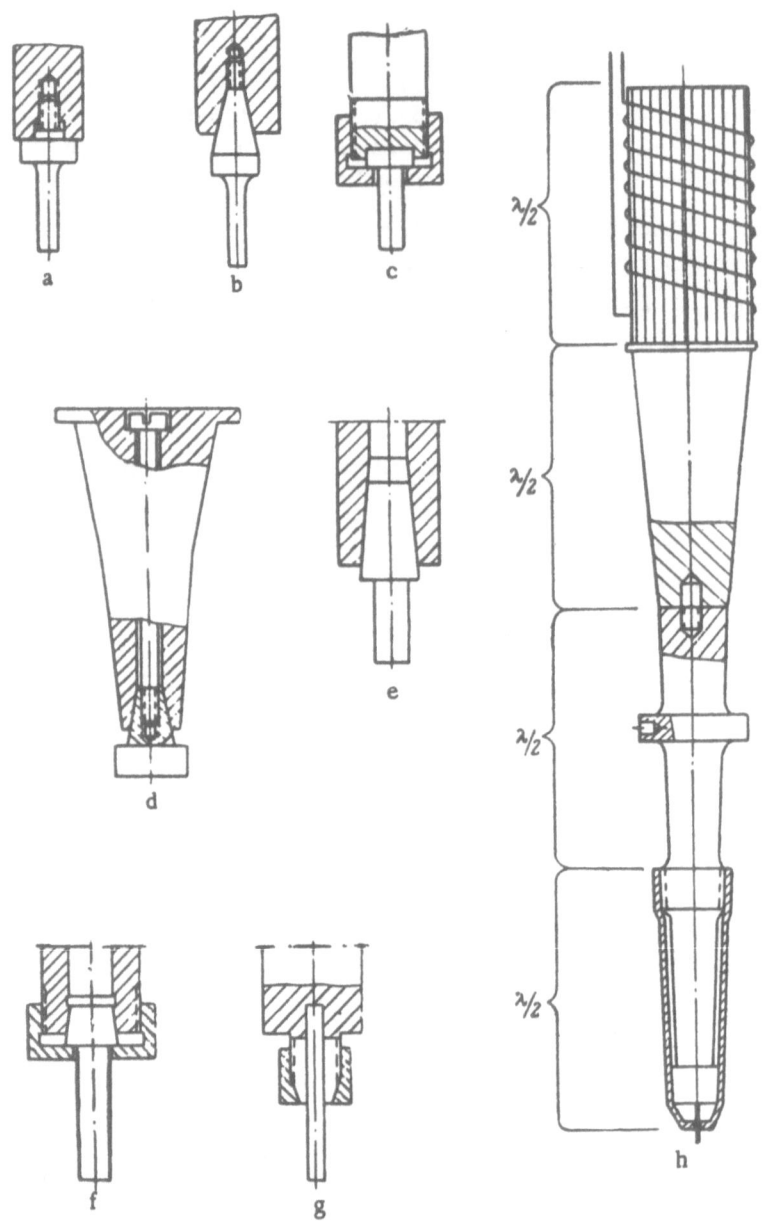

Fig. 127. Methods of attaching the tool: a) threaded shank, cylindrical part for centering; b) the same, but taper centering; c) shoulder and collar; d) drawbolt; e) simple taper; f) at node, with cone and collar; g) chuck at node [10]; h) chuck at antinode [16].

Table 22 [3, 17] indicates the regions of application of the various methods of attachment.

The design of tools, especially complicated ones, requires considerable experience.

<u>Abrasive Suspension</u>

The abrasive must be harder than the material, or at least of the same hardness. Abrasives in common use are boron carbide, silicon carbide, corundum, and (sometimes) diamond dust (Table 23). Boron carbide is the best as regards cutting properties, but it is in short supply and the most expensive (apart from diamond dust).

TABLE 22. Ranges of Application of the Various Methods of Attachment

Method	Permissible stress at join, kg/cm^2	Applications
Soft soldering or taper	80–110	Glass, ferrites, semiconductors
Thread	110–160	Ditto
Hard soldering	>160	Hard alloys, thermocorundum, hard ceramics

It is readily wetted by water and is of relatively low density, so it forms a satisfactory suspension. One disadvantage is that it contains many inclusions of graphite. Silicon carbide and carborundum are used on glass, germanium, and some ceramics. The use of silicon carbide in place of boron carbide increases the cutting time by 20-40%, while corundum requires a cutting time 3-4 times longer. Diamond dust is used in working diamonds and rubies.

A new promising abrasive is boron silicocarbide, whose abrasive power is 8-10% higher than that of boron carbide.

TABLE 23. Abrasives Used in Ultrasonic Machining

Abrasive	Composition	Ridgeway hardness	Density g/cm^3	Cost of 1 kg, rubles	Wettability
Electrocorundum	Al_2O_3 (1-2% Fe, chromium oxide),	12	3.93–4.0	Satisfactory	0.18–2.18
Silicon carbide (black)	SiC	13	3.15–3.20	Moderate	0.3–2.4
Boron carbide	B_4C	14	2.52	Good	8.0–10.3
Diamond	C	15	3.45–3.6	Not wetted	3 rubles per carat (0.2 g)

TABLE 24. Grain Sizes of Abrasive Powders

GOST 3647-59 number	Size, μ	Mesh	GOST 3647-59 number	Size, μ	Mesh
Grinding powders			Micropowders		
12	125–160	100	M 40	28–40	
10	100–125	120	M 28	20–28	
8	80–100	150	M 20	14–20	600
6	63–80	180	M 14	10–14	
5	50–63	230	M 10	7–10	
4	40–50	280	M 7	5–7	800
3	28–40	320	M 5	3–5	1000

TABLE 25. Fractional Composition (%) of Abrasive Powders (GOST 3647-47)

Grade	Larger grains not more than	Grains of main and next lower sizes not less than	Smaller grains not more than
Grinding powders	20–25	55–65	3–20
Micropowders	17–36	60–82	7–20

TABLE 26. Relative Ultrasonic Workability of Materials [19]

Material	Rel. workability, %	Ridgeway hardness
Glass	100	8-9
Germanium	200-250	7
Silicon	100-125	8
Ferrites	50	–
Quartz	50	–
Hard alloy (VK-8, VK-20) ...	2.0-2.5	13
Hardened steel	1.0	Up to 9

Grinding powders Nos. 6-10 are used in rough machining, and Nos. 3-5 in fine machining, or micropowders M28-M5 in finishing. Table 24 gives the grain sizes.

Each fraction contains not only grains of the main size but also larger and smaller grains (Table 25).

The carrier liquid also has the function of cooling the cutting area and flushing away the products. Water is the usual liquid, on account of its low viscosity, reasonably high density, satisfactory wetting, and good cooling.

However, water causes corrosion, so workpieces and machines must be protected. Machine parts may be made of stainless steel, and sodium nitride (0.5%) may be added to the water.

Micropowder in oil is sometimes used for fine machining, but the cutting rate is lower.

The abrasive is used at a concentration of 45-55% by weight for all grain sizes; this is reduced to 40-45% for pump feed, because the hoses tend to block at higher concentrations.

The grains are broken in the cutting zone, which tends to reduce the cutting rate. The resistance to breakage is governed by the hardness of the workpiece. A machine of medium power working on hard alloy uses 1 kg of boron carbide in 2-3 shifts. The relatively rapid reduction in grain size makes tool design difficult. Particles of cut material in the suspension have no major effect on the cutting [11].

Deposition in the feed system, rapid fracture, and evaporation of the liquid together make the actual abrasive concentration in the cutting zone rather indefinite. This explains the great spread in the experimental results.

Cutting Rate

The cutting rate* is governed by the impact brittleness and viscosity of the workpiece, other things being equal (Table 26).

The cutting rate is governed by many factors, of which the main ones are the amplitude and frequency of the vibrations, the hardness and grain size of the abrasive, the viscosity and abrasive content of the suspension, the cutting area, and the depth of cut. Chapter 2 deals in detail with these; here we consider only some practical points.

The amplitude is usually 0.02-0.06 mm, the cutting rate being roughly proportional to the square of the amplitude (i.e., to the power). The amplitude should be roughly equal to the grain size; the cutting rate falls rapidly for amplitudes below 0.02 mm [3, 18].

The maximal amplitude is governed by the fatigue strength of the concentrator; ξ_m is 60-70 μ for steel. A crack appearing at the end (antinode) of the concentrator soon grows and causes fracture; a crack at a node disrupts the coupling between the parts, so the resonance frequency falls, and with it the cutting rate.

The end of the tool is not always perpendicular to the axis; it is often of complicated shape, and may include inclined, concave, and convex parts. The component of the amplitude normal to the local surface is

---

* This may be specified as the volume cut (mm^3/min) or as the rate of penetration of the tool (mm/min).

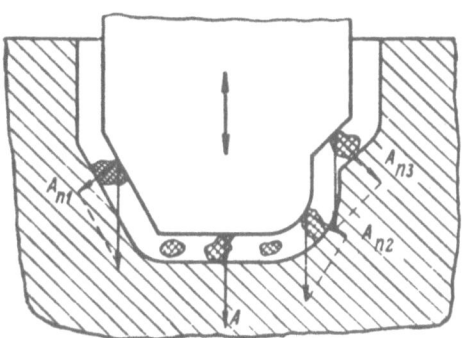

Fig. 128. Variation of the normal component of the amplitude for a tool of complicated end shape.

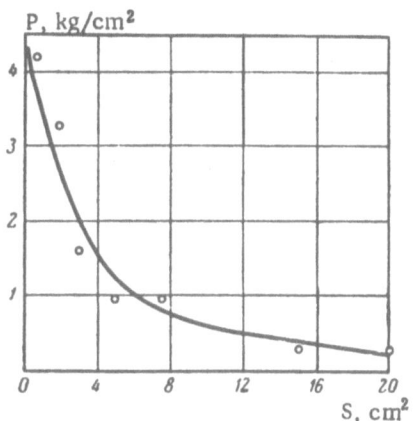

Fig. 129. Relation of optimal feed pressure to area.

governed by the angle between the surface and the axis, the relation being that of the sine of the angle (Fig. 128). The normal component determines the cutting rate, so this is lower on inclined areas, and the more so the greater the inclination. The tangential component plays no great part, for this produces only a low cutting rate. The concave parts are the least favorable as regards entry of fresh abrasive. The cutting conditions may thus vary over the end of the tool, and the over-all rate is governed by the parts where conditions are worst.

Lower working frequencies are better as regards the energy demand; moreover, the resonant system is then less sensitive to wear in the tool, which is especially pronounced in the machining of hard alloys. This aspect is considered in more detail below.

The rate is highest for a certain feed force, which is governed mainly by the cutting area. The maximum is more pronounced for small areas. The pressure is usually in the range 1-5 kg/cm² (Fig. 129), the higher values corresponding to finer abrasive [2, 3, 20].

It is sometimes stated that the probability of transverse vibrations in the tool increases with the feed force and that for any given conditions there is a definite range of feed force in which such vibrations are not excited [21] (Fig. 130). Of course, a very important factor here is the play in the slides of the feed mechanism.

The cutting rate also increases with the grain size (Table 27).

The rate is also much affected by the flow of abrasive to the cutting zone; the rate falls rapidly if the flow is impaired. Relative failure of this flow is responsible for the fall in cutting rate as the working area increases and as the tool penetrates, the relation being nearly exponential (Fig. 131). Various methods are applied to improve the flow: force feed, feed via service holes in the workpiece, periodic lifting of the tool (1-4 times a minute). In addition, flutes or holes in the tool may be employed. An effective method that produces a large rise in rate is to pump off the suspension via the tool (Table 28). The rate then varies little as the tool penetrates (Fig. 132).

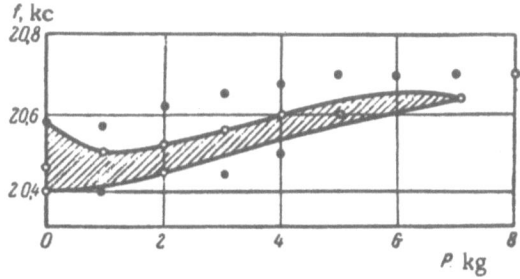

Fig. 130. Hatched region denotes transverse vibrations absent, the frequency range being that for resonance in the system [21].

Accuracy and Finish

The accuracy is governed mainly by the grain size of the abrasive (Table 30); the influx of abrasive along the sides causes the hole to be larger than the tool, the clearance being larger at the entrance than at the exit (Fig. 133). The reason is that the grains are largest as they enter at the top and tend to be broken as they migrate downwards. Moreover, transverse vibrations are more likely to occur at the start of a cut.

TABLE 27. Highest Cutting Rate (mm³/min) for Hard
Alloys at the Start of Cutting for Various Grain Sizes [13]

Power, kW	Grain size (boron carbide)		
	10	5	3
0.25	6	4	2
1.0	16	12	6
2.0-3.0	36	25	12

The lateral clearance is roughly one and a half times the grain size. For instance, hard alloy worked with No. 10 abrasive (diameter 105-125 μ) gives a clearance of 0.22-0.28 mm on the diameter, whereas No. 3 (diameter 28-42 μ) gives 0.15 mm clearance [9].

This clearance means that the tool must be smaller than the required hole size in order to provide the specified tolerance. Table 29 gives the reduction in size needed on the assumption that the hole is made in one pass by a single tool.

Wear in the tool introduces an additional effect and increases the angle of the hole, while sharp corners become rounded (Fig. 133). This means that tool replacement is essential in the production of accurate blind holes; the rough machining operation must be followed by a fine cut, and sometimes also by a finishing cut.

The grain size δ_2 that should be used for the fine cut may be calculated from

$$\delta_2 = \frac{(a - d_2) - 2(\Delta_1 - K_1\delta_1)}{2K_2}, \qquad (5.3)$$

in which a is the specified size, d_2 is the diameter of the tool for this cut, Δ_1 is the clearance during rough cutting, δ_1 is the grain size used in rough cutting, and K_1 and K_2 are factors equal to the ratio of the clearance to the mean grain size ($K_1 \approx K_2 \approx 1.6$) [23].

Each new tool must be 0.02-0.03 mm larger than the previous in order to leave no steps on the surface.

Fine machining of through holes may be done with the unworn part of the tool for 10-20 min, the tool reciprocating and the suspension being fed in. The cone angle can be reduced to 2-3' in this way.

One way of reducing the taper of a through hole is to cut it in turn from both sides, the first hole serving as basis for setting the workpiece for the second [24].

Transverse vibration in the tool (on account of incorrect design or incorrect setting) causes an enlarged hole, particularly at the start; it may also cause chipping at the edges and uneven wear on the tool. Careful attention must be given to suppressing such vibration.

The machining error is governed by the precision of the machine, especially as regards the acoustic head and feed mechanism. It is also governed by the workpiece material. For instance, it is difficult to

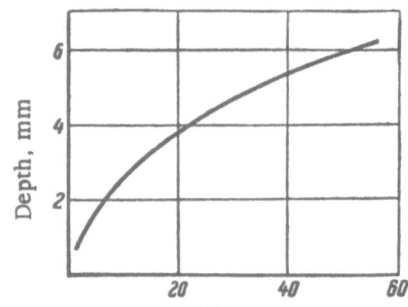

Fig. 131. Cutting time as a function of depth of tool.

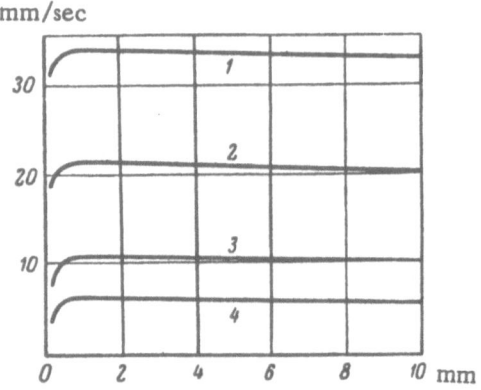

Fig. 132. Cutting rate as a function of depth when abrasive suspension is pumped off. Hollow tool 17 mm in diameter, amplitude 30 μ, 20.7-21 kc, feed force 4 kg, 25% boron carbide No. 320 in water, plate glass. 1, 2, 3, and 4 relate to tools with central holes of diameters 3, 9, 13, and 15 mm respectively [21].

TABLE 28. Effects of Withdrawal of Suspension on the Cutting Rate [22] (22 kc, 300 W, boron carbide No. 280, silver-steel tube tool, outside diameter 10 mm, inside diameter 8 mm, depth of hole 5 mm without pump-off and 20 mm with pump-off)

Material	Rate, mm^3/min	
	no pump-off	with pump-off
Glass	200	1000
Ferrites	150	800
Germanium	180	800
Carbon, graphite	100	500
Quartz	100	400
Silicon	80	400
Agate	60	320
Hard ceramic	50	250
Mother-of-pearl	50	250
Ruby, sapphire	8	30
Hard alloy	5-7	30-40
Boron carbide	1-4	2.5-10

machine a material of highly granular structure or large grain size (e.g., corundum, silicon carbide, ferrites) with high precision, because large particles become detached; for example, it is difficult to work a ceramic to closer than 0.05 mm.

The surface finish is governed by the grain size of the abrasive (Table 31) and also by the material of the workpiece.

Cavitation erosion causes the surface finish of the sides to be lower by 1-2 classes than that of the end face. The cutting should be performed at the highest possible speed (with optimum values of the parameters) in order to minimize cavitation effects. Cavitation also damages the tool; it produces pits on the end, which are reproduced in the workpiece. The surface finish is improved by periodic regrinding of the end. Cavitation erosion occurs on the sides and end of tool and workpiece.

Softer or harder inclusions in the material or tool may cause local defects.

The ultrasonic method represents the finishing operation for many components for which classes 1-3 of precision are acceptable; higher precision is difficult to ensure, on account of the large number of parameters involved.

Ultrasonic broaching has been considered above. Ultrasonic surface grinding can give high precision; for instance, variations in surface height of 2-4 μ on hard alloy can be reduced to 0.2 μ in a few minutes.

TABLE 29. Accuracy and Finish as Affected by Grain Size of Abrasive for Workpieces of Hard Alloys*

Abrasive No.	Grain size, μ	Limits, mm†	Cone angle	Finish of sides	Finish of end face	Reduction in tool size, mm	Reduction for tools to machine holes of angle 1°30', mm
10	105-125	± 0.04	3°	▽5-▽6	▽ 7	0.33	0.17
5	63-75	± 0.02	1°45'	▽6	▽ 8	0.24	0.12
3	28-42	± 0.01	1°15'	▽7	▽ 9	0.12	0.05
M14	10-14	± 0.005±0.007	40-50'	▽8	▽10	0.07	0.03

* The surface finish is poorer by 1-2 classes in the case of glass.
† Limits are for clear holes; they are lower by about a factor 2 for blind holes.

TABLE 30. Relation of Maximum Area to Power Available [27]

Power, kW	0.1	0.25	0.5	1.0	2.0
Maximum area, mm^2 ..	20-80	80-180	180-500	500-1000	1000-2400
Area of end of transducer, cm^2 ..	2	5	10	20	40
Maximum diameter of solid tool, mm ...	5-10	10-15	15-25	25-35	35-55

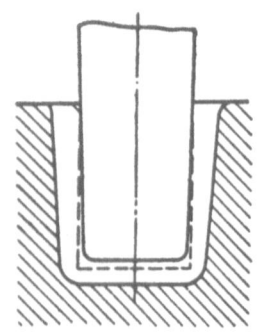

Fig. 133. Wear of tool and distortion of hole in ultrasonic drilling.

An example of such grinding is as follows. A tool of diameter 15 mm may be made of mild steel and may be driven with an amplitude of 20 μ at 28 kc. The workpiece may be rotated at 100-800 rpm around its own axis and 6-100 rpm around the axis of the table (combination of motions), the pressure being 30-60 g/mm^2 [25]. Similar conditions have been used in the ultrasonic grinding of glass [26].

Allowance must be made for the overlap between tool and workpiece; incorrect choice of this can result in a raised or rounded edge [69].

Size of Holes

Machines of output power 0.05 to 2.4 kW so far made are capable of handling holes of diameters from 0.075 to 90 mm, the maximum diameter being governed by the power (Table 30) and the minimum by the strength of the tool and the sensitivity of the feed mechanism. Further increase in size requires a larger working surface in the transducer, which involves certain difficulties.

The maximum drilling depth is 2-5 times the diameter of the tool and is governed by the conditions for abrasive supply. Special measures to ensure good supply (e.g., periodic withdrawal of tool or special systems) enable one to increase the cutting depth.

The length of the tool must be at least twice the thickness of the workpiece in drilling hard alloys. The change in length as a result of wear should not exceed $\lambda/8$ [see (5.2)], so the maximum thickness that can be penetrated is

$$ h = \frac{\Delta L}{2} = \frac{\lambda}{16} = \frac{c}{16f}. \tag{5.4} $$

This means that thicker workpieces can be accommodated at lower frequencies.

Effects of Ultrasonic Machining on the Material

The total cutting force is small, so the process gives rise to no appreciable mechanical stresses sufficient to cause warping or other residual deformation. The heating is negligible if the process is operated properly, so the structure is not affected. However, strong heating can occur if the flow of suspension is impaired, which can result in cracking.

Chipping may occur at the exit side in drilling through holes, because the brittle material starts to work in bending. This may be overcome by fastening the workpiece to a base (usually of glass) with wax, picein, or a wax-rosin mixture (1:1). A steel base may sometimes be used, as this can be reground as required. Another method of eliminating chipping is to reduce the feed force as the exit is approached.

Combination of Ultrasonic and Electrical Methods

The electrical methods are applicable to conducting materials, so the two can be combined for hard, refractory, or magnetic alloys, and also for hardened steels. The two methods may be used in sequence or simultaneously.

The Italian firm of Federici has produced a combined ultrasonic and electroerosion machine, whose output power for electroerosion is 0.75-6 kW (ultrasonic 1.6 kW); the two methods can be used simultaneously or separately [28].

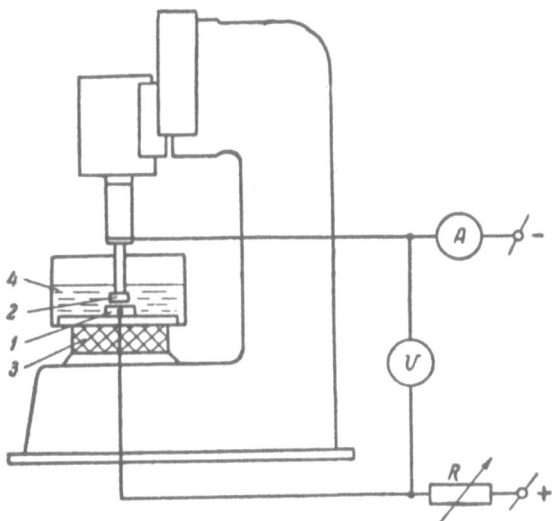

Fig. 134. Machine for combined ultrasonic
and electrochemical machining.

Sequential use is mainly employed for hard alloys and rather less often for hardened steel.

Electroerosion allows one to perform the same operations as ultrasonics in the case of hard alloys, but the surface finish and precision of ultrasonic machining are better.* The more rapid electroerosion method is therefore used for rough machining; the residual material (0.1-1 mm) is removed with ultrasonics, to give high accuracy and good surface finish † [11, 13].

It is found that the ultrasonic vibrations stabilize the impulse machining when the pulses are passed through the tool; the cutting rate is increased and the tool wear is reduced. For instance, steel 45 worked at pulse rates of 400-1500 per sec is then cut at more than double the rate, while the tool wear is reduced to a third [29].

The vibration favors deionization in the discharge gap in electrospark working. The process remains stable at higher working currents, which means higher cutting rates. Tool wear is somewhat reduced [30].

Combination of ultrasonic and electrochemical methods provides an increase in cutting rate on hard alloy by nearly a factor of 3, and tool wear is reduced. In this case, ordinary salt (20-30%) is added to the suspension, the current density at the tool (cathode) being 15 A/cm². It is considered that the electrochemical process converts the metal to the brittle oxide, which is readily cut by the ultrasonic process. Further, the ultrasonic cutting accelerates the electrochemical attack. Figure 134 shows a machine for combined use of these methods [31, 32].

Economics and Practical Features

Ultrasonic machining has the advantage of providing a means of machining hard and brittle materials to complex shapes with good surface finish and accuracy. On the other hand, there are numerous disadvantages, mainly because there has not been enough research on the techniques and machines.

The method is less rapid than diamond machining in the case of cylindrical surfaces; it is used only when the latter equipment is lacking in small-scale production. It is also unsuitable for use with ordinary constructional materials.

Considerable economy results from ultrasonic machining of hard-alloy press tools, dies, and wire-drawing equipment, on account of the high wear resistance of tools made of these alloys (16-100 times that of steel

---

* See also section 26.

† It has been found that the altered surface layer resulting from electroerosion is readily worked by ultrasonics.

ones). Moreover, the method accelerates the actual production of these dies by factors of 4 to 25. The tolerances can be reduced to ± 0.01 mm. The cost of hard-alloy wire-drawing dies is about 5 times greater if made in the traditional way.

The saving from introducing one ultrasonic machine for the production of hard-alloy tools is 30 thousand rubles a year (one such press tool replaces 20-40 steel ones).

There is a saving of 20-50% on material in the case of parts made of glass, quartz, or fluorite. The cutting rate is increased by a factor of 8-10. The annual saving per machine in this case is 10 thousand rubles (6.3 thousand for the piercing of precious stones).

With brittle materials such as ferrites, germanium, and silicon the method increases cutting rates by factors of 5-8 and also reduces the wastage, on account of the reduced hazard of chipping during drilling.

The power consumption of ultrasonic machining is 0.1 kW-hr/cm^3 on current machines for glass, or 5 kW-hr/cm^3 for hard alloys.

The main costs in operating an ultrasonic machine come in making and using the tools, especially if these are of complex shape.

Abrasive is a major item in the costs. The grains are eventually broken and blunted, and the suspension becomes depleted on account of deposition and removal together with finished workpieces. The suspension must be periodically replaced. The type of abrasive is chosen to suit the material. Boron carbide is expensive and should not be used in any case in which corundum will give good results.

Ultrasonic machines are not yet sufficiently reliable; they sometimes fail on account of faults in acoustic heads, pumps, and generators.

The heads fail from fracture in parts of the vibrating system, which work under high sign-varying loads and which may be subject to cavitation erosion. More serious damage may occur if joints fail or the transducer becomes deformed.

The main reason for pump failure is entry of abrasive into the electric motor.

Vacuum tubes have only a limited working life, so the generators need periodic skilled servicing.

Ultrasonic cutting began to be introduced into industry in 1951-1953, but the techniques are still not properly worked out. It is still sometimes necessary to do a series of trials to find the proper dimensions for tools when starting the production of some new component.

The machines have no parts moving at high speed; contact with the tool does not involve a hazard of injury to the worker. The sides of a vibrating tool feel slippery or velvety to the hand; touching the end produces a painful sensation resembling a burn, especially if the power is high.

No special measures to protect the user from the ultrasonic emission are taken in current machines, for the tendency to radiate is slight, and the radiation is quite strongly absorbed. On the other hand, high-intensity ultrasonic radiation is hazardous and should not be allowed to fall on any part of the body.

26. Industrial Use of Ultrasonic Machining
Range of Application

The method is already used in, or appears very promising for:

1) making press tools, dies, wire-drawing equipment, and other components consisting of hard alloys;
2) producing metal-working tools fitted with hard-alloy or cermet tips;
3) fabricating components of glass and quartz;
4) forming semiconductor (germanium and silicon) components;
5) finishing cermet radio and electronic components;
6) working ferrites;
7) working diamonds;
8) machining technical stones and jewels.

In addition, ultrasonic cutting is applied in dentistry and in engraving processes.

The most important current industrial developments are the manufacture of hard-alloy dies and wire-drawing equipment, and the working of germanium and silicon, optical glass, ceramics, and diamonds.

The above list is far from exhausting the possible applications.

For instance, ultrasonic methods may be used in sharpening tools bearing hard-alloy or cermet tips.

Ultrasonic Machining of Tools and Other Hard-Alloy Components

Hard alloys can be machined by various electrical and other methods as well as by ultrasonic techniques [3, 9, 11, 13, 15, 33]: electrospark, electric pulse, anode-mechanical, electrochemical, and diamond grinding. Each method has its advantages and disadvantages; methods should be chosen in accordance with the detailed nature of the problem.

In any case, it is best to use hard alloys in their soft state, as this greatly reduces the volume of subsequent machining.

The ultrasonic method in its present form is suitable for use on surfaces of comparatively small size. It is necessary to use several passes or to apply entirely different machining methods if the area is large.

Components with long and narrow slots are best machined by the wire electrospark method [34], since thin ultrasonic tools wear rapidly and are very susceptible to transverse vibration, which increases the cutting error.

Electrical methods provide high output in rough machining,* whereas the ultrasonic method gives better surface finish. The preliminary machining is therefore best done with one of the electrical methods, leaving 0.1 to 1 mm to be removed by the ultrasonic method to give a good finish. The holes and so on produced by the rough machining have the incidental advantage of facilitating access by the suspension.

On the other hand, it is best to use ultrasonic cutting throughout for small components.

The rapid tool wear on hard alloys makes it best to use two ultrasonic cuts. The first cut is performed with a tool 0.2-0.4 mm undersize, with No. 6-10 boron carbide; the second (finishing) cut is made with a tool 0.08-0.16 mm undersize and No.3 abrasive (or even a finer grade).

It is common to find that a tool is good for only one workpiece in the case of hard alloys.

On the other hand, a tool that has been used for the finishing operation can then be used for the rough machining. If possible, the tool should be hollow, with a wall thickness not less than 1.5-2 mm (for strength) and of length 25-30 mm.

Ultrasonic machining is used in making hard-alloy tools for upsetting, deep drawing, and extrusion forming; the alloys are grades VK8-V, VK-15, VK-20, VK-25, and VK-30.† The method is best used on integral dies of complex shape.

The method is also applicable to forging and stamping tools.

Ultrasonic cutting usually provides adequate accuracy and surface finish in hard-alloy dies; it therefore can form the finishing operation.

An upsetting or drawing die is made as follows:
1) a through hole is formed by electroerosion, leaving 0.1-1 mm to be removed by ultrasonic cutting;
2) a first ultrasonic cut is taken with No. 5-6 boron carbide as abrasive;
3) the second ultrasonic cut is taken from the other side, with boron carbide No. 3 or M10;
4) honing may be applied to remove any slight taper.

This method gives dies accurate to 0.01-0.02 mm; the operations take 0.5 to 1.5 h (for apertures 20-50 mm in diameter (Table 31).

---

*Pumping-off of abrasive increases the cutting rate of the ultrasonic method considerably, but it cannot always be used.

†Hard alloys containing cobalt are cut at roughly constant rates no matter what the cobalt content.

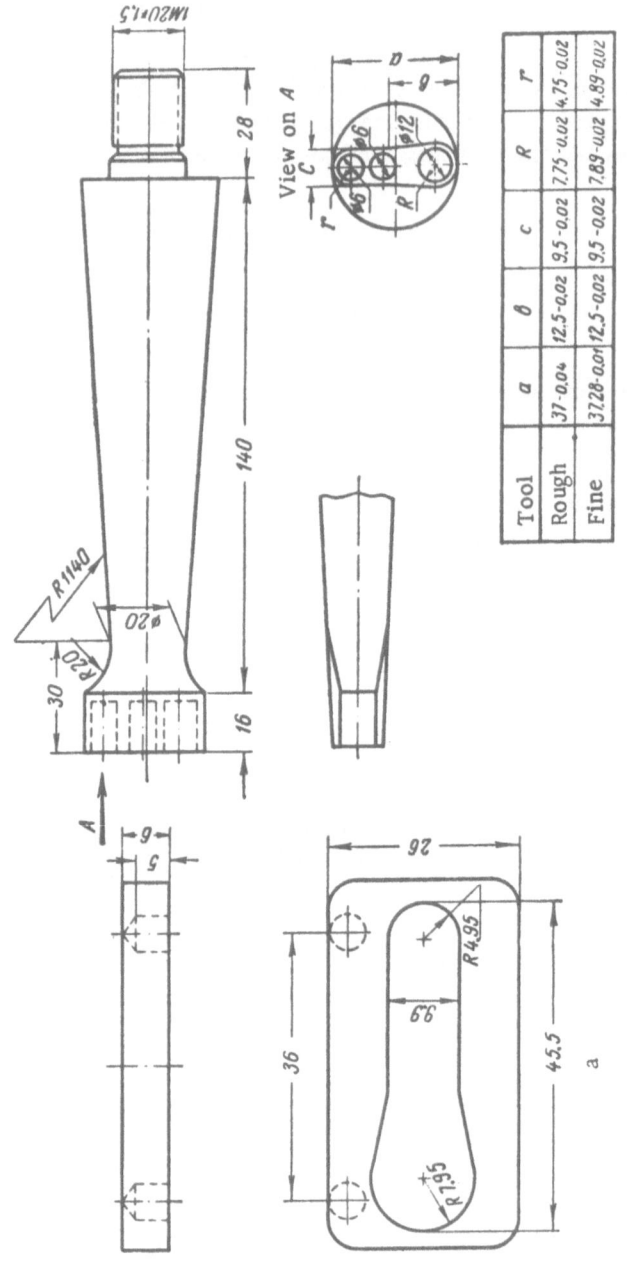

Tool	a	b	c	R	r
Rough	37 -0.04	12.5 -0.02	9.5 -0.02	7.75 -0.02	4.75 -0.02
Fine	37.28 -0.04	12.5 -0.02	9.5 -0.02	7.89 -0.02	4.89 -0.02

Fig. 135. a) Hard–alloy insert for a die for stamping springs; b) tool used to form die [9].

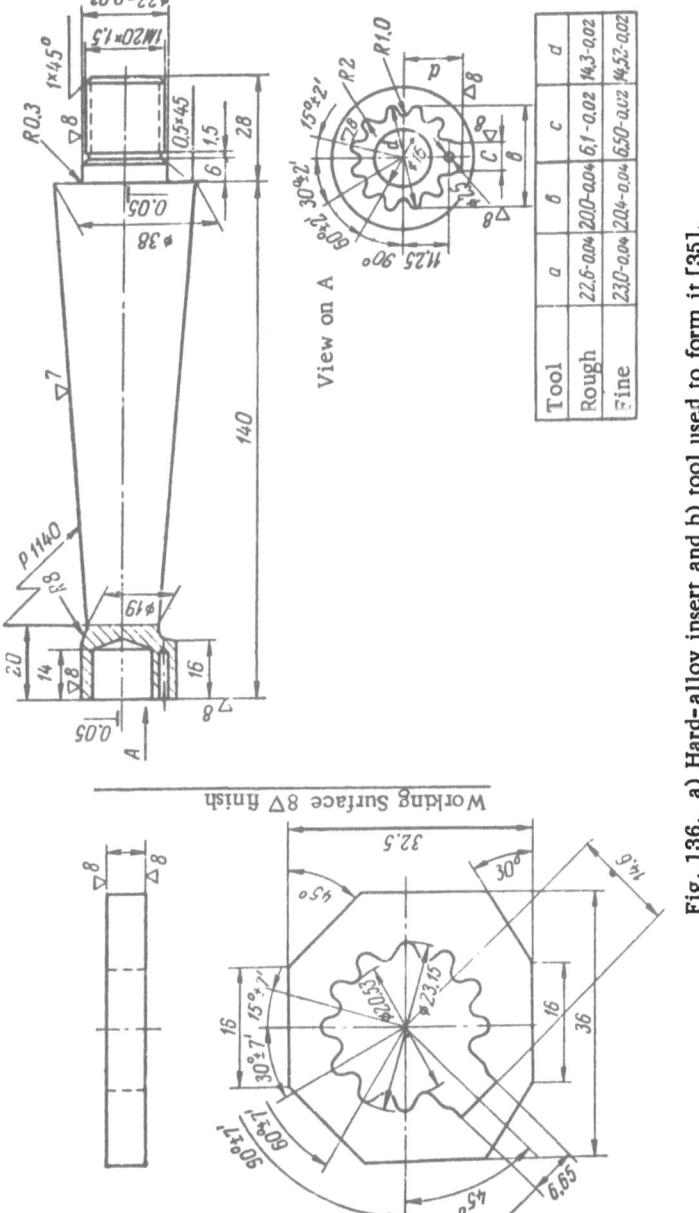

View on A

Tool	a	b	c	d
Rough	22.6-0.04	20.0-0.04	6.1-0.02	14.3-0.02
Fine	23.0-0.04	20.4-0.04	6.50-0.02	14.52-0.02

Working Surface ▽8 finish

Fig. 136. a) Hard-alloy insert and b) tool used to form it [35].

129

TABLE 31. Time Taken by Various Operations on Hard- Alloy Components [12]*
(Machine Power 1-3 kW)

Hole	Size, mm	Thickness, mm	Time, min	GOST 3238-46† grade of abrasive
Circle	φ3	4.5	15	5
	φ7	4.5	20	5
	φ15	6	40	5
	φ20	20	30	M14
	φ32	20	12	M14
	φ47.5	5	60	5
	φ76	4	340	M14
Square	3 × 3	4.5	15	5
	6 × 6	4.5	30	5
Slot	0.5×16	4.5	20	5
	0.5×20	6	50	M14
	1 × 14	6	30	5
	1 × 40	6	60	M14
Rectangle	10 × 40	6	90	M14
	28 × 65	3	230	M14
E shape	30 × 45	6	200	M14
15 holes	φ5 × 15	6	212	5
Star	φ28	14	68	8
Ridges	20-30	5	30	5

* Hole made in blank by electroerosion, leaving 1 mm for ultrasonic cutting.
†M14 micropowder is used for fine machining after rough machining with No. 5-6.

A shallow punch or drawing die may have a recess 10-15 mm deep and of taper 30-90'. Such a hole may be formed in one pass, in which case the working part of the tool must have a length 3-3.5 times the depth of the hole, the diameter being reduced in accordance with the grain size (section 25).

It is best to form a hole of special shape by one pass of an appropriate tool, although it is possible to use several simple tools in succession, especially if slots, shoulders, and so on have to be formed.

Round holes in dies are best cut to a depth of 10 mm under the conditions of short-run production or if the equipment for abrasive grinding and honing is lacking.

Figures 135 and 136 show dies made of VK8-V alloy; each took about one hour to make, the cutting being done in two passes: first a rough cut with No. 10 boron carbide and then a finishing cut with No. 3 boron carbide.

Fig. 137. Dies of special profile cut from hard alloy by ultrasonic machining [36].

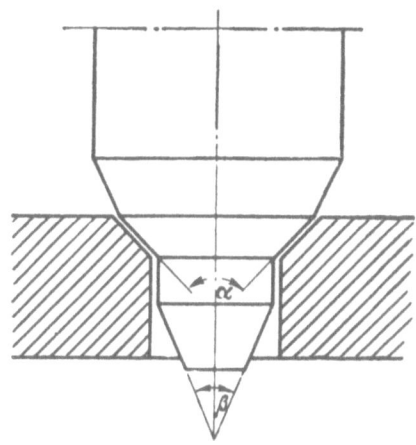

Fig. 138. Machining of shoulders and wear on tool.

The accuracy is of class 2-3. The holes at the ends of the tools serve to minimize the volume of material to be cut and to ensure even wear over the end of the tool [9,35]. Figure 137 shows dies of rather more complex shape, also made in the same way.

It is sometimes best to use a hard-alloy die with a steel punch; then the punch and the cutting tool can be made simultaneously from a single piece of steel, which is cut in half, one piece serving as the cutting tool and the other as the punch. Any necessary size reduction can be produced by etching [37].

Drawing dies are made in much the same way as punching ones, by drilling holes, but the corners are then rounded off.

A conical tool is used to remove shoulders; the angle α of the tool must be less than that of the finished shoulders (Fig. 138). Veroman's formula for the angle of the cone is

$$(1 + K)\cot\alpha = \cot\beta, \qquad (5.5)$$

in which K represents the relative wear of the tool.

It is difficult to take detailed account in advance of tool wear for rounded areas, so it is best to perform the operation in several steps with new tools (or with a reground tool) in order to approach the correct shape.

Forging and stamping dies, and punches having three-dimensional surfaces, are more difficult to make, because end wear on the tool is more important here.

Such components are often made by using several tools in turn. The number of tools or regrinding cycles for a final tolerance of 0.05 mm is

$$n = \frac{3 + \ln H}{0.7}, \qquad (5.6)$$

in which H is the depth of the hole in mm [11].

Dies for embossing are made by the use of a tool engraved at the end in the usual way. Deep engraving presents difficulty, so it is better to use a fresh tool at each stage rather than regrind. This requires 4 to 7 tool changes.

Ultrasonic machining is particularly efficient if the tools are simply items embossed in such a die. For instance, the dies for pressing the characters of typewriters and accounting machines are made by the use of such characters. These dies are of small depth and area, so the ultrasonic cutting operation takes only 96 sec. The operation is done in two passes on two machines [38].

The techniques of making hard-alloy dies for extrusion and wire-drawing are much the same as for other dies, but the final polishing is performed by hand (Fig. 139). The closest limits attainable are ±7.5 μ. It is best to confine ultrasonic machining to dies of complex shape (Fig. 140), since ordinary mechanical methods are quite satisfactory for holes of round, hexagonal, and other simple shapes. Figure 141 shows a die formed by ultrasonic machining.

Ultrasonic cutting is also used in making hard-alloy gages (Fig. 142) and control components for aircraft hydraulic systems; the advantage of the method for pneumatic gaging tools is that the gage parts can be integral, as no soldering operation is needed, which is the main cause of scrap [39].

Manufacture of Glass and Quartz Items. Engraving.

Ultrasonic methods are particularly efficient in the working of glass, especially for the cutting of blanks for optical components, drilling, and engraving.

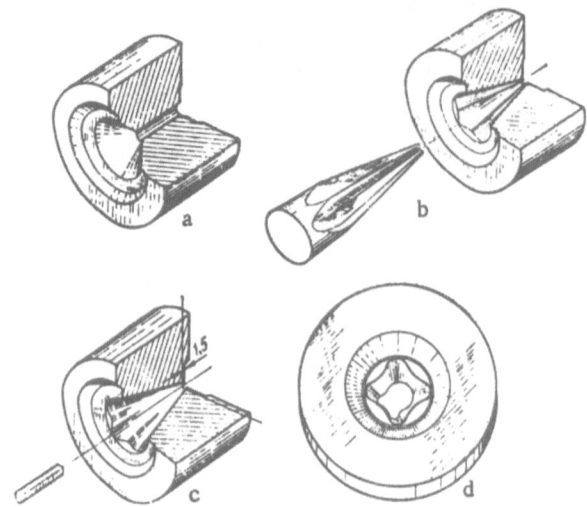

Fig. 139. Sequence of operations forming a profile wire-drawing die: a) blank; b) formation of cone with shaped tool; c) finish cutting of exit section (boron carbide No. 320, time 5 min); d) finished die after polishing. Limits ± 0.0075 mm.

Fig. 140. Profile extrusion die made by ultrasonic machining.

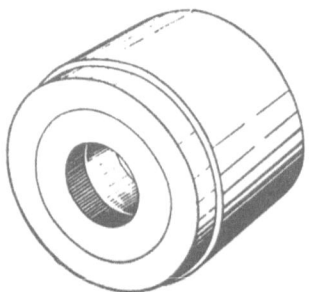

Fig. 141. Die of hard alloy formed by ultrasonic machining.

Fig. 142. Tip of measuring instrument with hard-alloy inserts having holes broached by ultrasonic machining: top, demountable insert made by grinding; bottom left, one made by ultrasonic machining.

The efficiency is most obvious for components of complex shape. For instance, the use of ultrasonic cutting for slots such as those shown in Fig. 148 reduced the cutting time from 30 to 1.2-2 min (thickness 6 mm, boron carbide No. 5); again, the cutting time for blanks for the lens of Fig. 143 was reduced from 6 h to 1 min, and less highly skilled workers were needed; in addition, there was a great saving on rejects (12 blanks cut simultaneously, with No. 3 boron carbide) [40].

Ultrasonic drilling of holes less than 8-10 mm in diameter increased labor productivity by a factor of 2-3 relative to ordinary drilling. Holes in small glass components can be cut to class 7 limits when abrasive of grain size No. 5 is used (Fig. 144). The machine time for drilling holes $\phi 2A_7$ and $\phi 6.5A_7$ was 0.4 and 1.1 min, respectively.

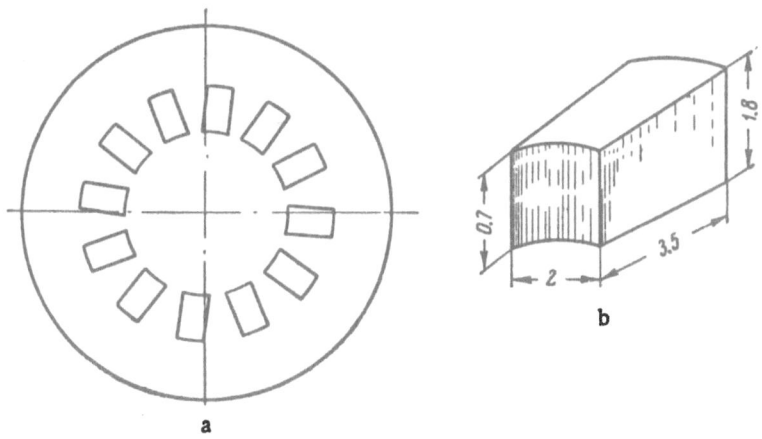

Fig. 143. Cutting of blanks for glass prisms: a) plate
with prisms cut out; b) prism.

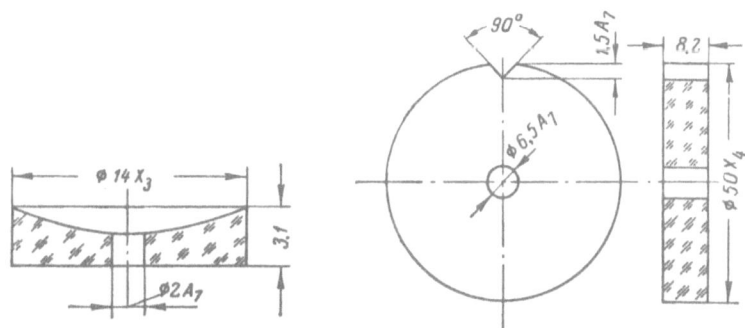

Fig. 144. Glass components having holes drilled
with ultrasonics.

Fig. 145. Ultrasonically engraved glass.

Ultrasonic cutting has been applied to cutting lens blanks; the tubular tool had a cutting rate of 7 mm/min. Blanks 5 mm in diameter and 5 mm thick took only 5-7 sec to cut with boron carbide No. 5 as abrasive.

The limits for ultrasonic machining of glass are usually not closer than ± 0.05 mm, but rotation of the workpiece during broaching enables one to reduce the limits to a few microns (tests have been done with a tool 8 mm in diameter on plate glass 6 mm thick with 80-800 mesh boron carbide at 28.8 kc, rotation speed 33 rpm) [41].

An ultrasonically machined surface has a matt finish; a workpiece with a polished surface should be coated with lacquer to protect it from the abrasive.

The most frequent cause of rejection with glass components is chipping at the ends of drilled holes. Ones at the entrance side are usually caused by transverse vibration of the tool or by blows on the very largest grains

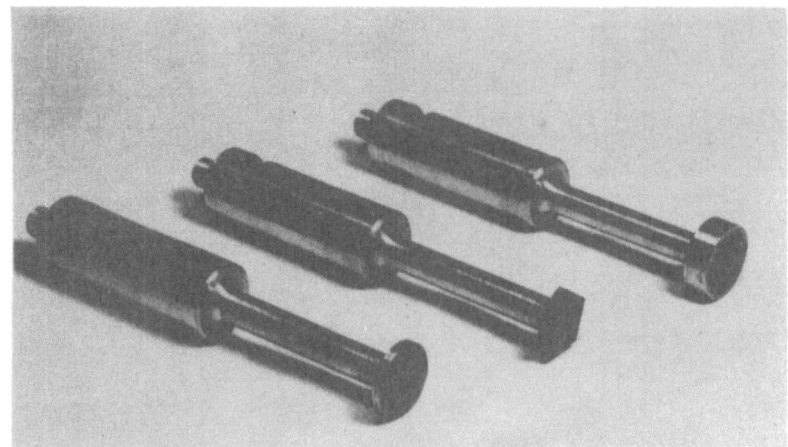

Fig. 146. Engraving tools.

of abrasive. These can be prevented by suppressing such vibrations and by using carefully graded abrasive. Another method is to use a slip of glass glued to the workpiece [70].

Chipping at the exit side is suppressed by cementing the workpiece to a glass base or by reducing the force as the exit is approached.

Ultrasonic engraving of glass is particularly effective.

There are three methods of ultrasonic engraving:
1) the pattern is produced by a tool bearing a combination of shallow (0.2-0.4 mm) lines with uncut areas, which are polished (Fig. 145);
2) the lines are engraved with a general tool;
3) deep engraving.

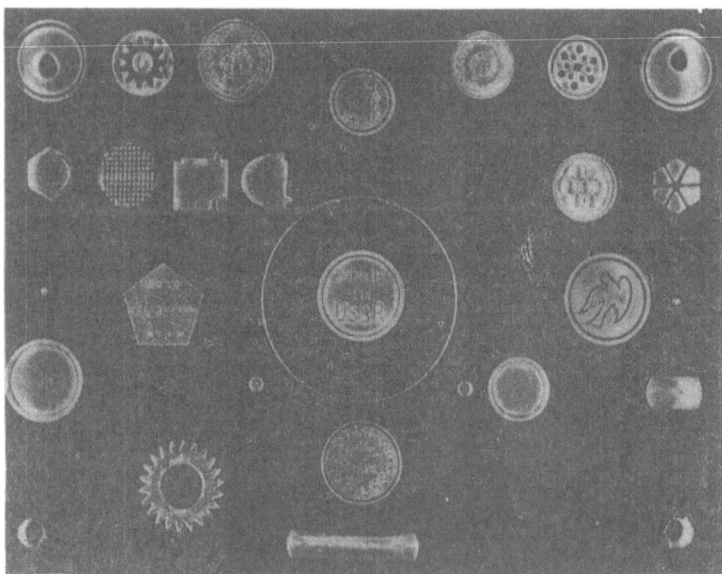

Fig. 147. Mirrors with ultrasonic engraving.

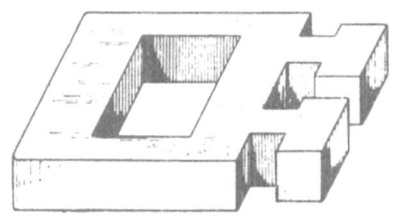

Fig. 148. Quartz cut to a special shape.

In the first case the tool has the picture engraved on its end to a depth of 0.5-1.5 mm (Fig. 146); it is applied to the entire surface.

The area (diameter 15-25 mm) is engraved in 2-5 sec by a machine of power 0.25 kW; the appearance is particularly striking if the engraving is done on a mirror (Fig. 147). The method has the disadvantage of small scale and the use of expensive tools (one tool is good only for 100-300 operations).

A three-position automatic machine employing this method gives an output of 550 components per hour engraved with patterns [42].

In the second case the pattern is engraved with a manual ultrasonic head. The abrasive suspension is applied by hand to the surface, the pattern being traced out with a pointed tool along the lines of a stencil. The engraving depth is only 0.2 mm, and the speed is very much that of drawing in pencil. This method is used for engraving pictures and inscriptions on glass vessels.

Deep ultrasonic engraving is used on gemstones; engraving to depths of 3-6 mm over areas 40-120 mm in diameter can be performed in 20-150 min. A steel tool lasts for 25 operations on average; a brass one, for 23.

Ultrasonic profile cutting, drilling, and slicing are applied to quartz as well as glass.

Figure 148 shows a profile-cut component of quartz having a window 35 × 20 mm cut in this way. This operation was performed with a knife tool 1 mm thick, the window being trepanned out in four cuts [40].

Holes 6 mm in diameter and 13-14 mm deep can be drilled at the rate of 20 an hour in fused quartz; this is somewhat more than in mechanical (rotating tube) drilling, though the energy consumption is higher. The tool was a steel tube of wall thickness 0.5-0.7 mm, the abrasive being No. 8 green silicon carbide [43].

A multiblade tool is used to slice quartz. For example, 21 plates each 15.9 × 15.9 mm or 19 × 31.7 mm (thickness 0.305 mm) were cut by such a tool in 10-25 min on a machine of power 0.4 kW with a vibration amplitude of 12.5 μ. Ultrasonic cutting produces twice as many plates from a given blank as does slicing with a diamond disc [44, 45].

A single-bladed tool does not give good results; the workpiece often breaks.

Machining of Semiconductors

Germanium and silicon are hard and extremely brittle, so many attempts have been made to apply ultrasonic methods to slicing, cutting blanks, and preparing other parts.

Ultrasonic methods have so far not been applied to the industrial cutting of monocrystals, although it has been shown possible to cut crystals 13-32 mm in diameter into wafers not less than 0.37 mm thick, with a production rate of 120 per minute [46, 47].

Ultrasonic cutting of round blanks from germanium and silicon plates improved output by factors of 5-8, reduced the proportion of rejects, greatly reduced the scrap, and increased the accuracy. Slicing into a mosaic is performed with a multibladed tool consisting of many tubes or plates, or of a solid rod having many holes drilled in the end; these tools can cut several hundred wafers simultaneously and give an output of 11,000 pieces an hour [48, 49].

Hollow tools are used to cut large-diameter discs of germanium and silicon. A disc 8 mm in diameter and 1 mm thick is cut in 15 sec, while ones 30-50 mm in diameter and 0.8-1 mm thick are cut in 30-60 sec. The abrasive may be electrocorundum or boron carbide.

One tool is good for 300-400 blanks in cutting semiconductor plates [4]. Cracking as the tool breaks through must be suppressed with particular care. The following method is sometimes employed to prevent breakage.

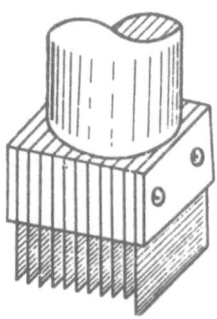

Fig. 149. Tool for cutting a mosaic.

Fig. 150. Mosaic cut from a plate
of semiconductor by ultrasonic machining.

The blank is cemented to a base and is cut almost completely through. Then a second plate is cemented on top, the bottom plate is removed, and the material holding the largely cut parts is machined away [3].

The multibladed tool of Fig. 149 is used to cut mosaics. The steel blades are 0.127 mm thick and have a pitch of 1.14 mm. The blade at the edge projects an extra 0.5 mm and is used to set the workpiece relative to the tool. The operation is conducted as follows. The semiconductor plate is cemented with shellac to a ceramic plate, which has two slots, lying at 90°, to guide the edge blade.

The base plate is clamped to a table having a ball joint, which facilitates correct setting; a series of slots is cut to a specified depth. Then the workpiece is turned through 90° and fresh slots are cut. The result is square areas 1.016 × 1.016 mm on the plate.*

These slots may be cut with two sets of blades differing in pitch to give areas of rectangular shape (Fig. 150). Powers of up to 0.5 kW are needed for a multibladed tool 22 × 22 mm.

The surfaces shown in Fig. 151 are generated by first cutting slots and then using special tools; specimen 1 was produced with a tubular tool, and 2 with a tool of square cross section having a hole at the end. In this way projections 0.05-0.305 mm in diameter have been generated, with cutting rates of 0.152-0.763 mm/min.

Powers of only about 60 W suffice for such operations. The machine needs to have a fine control for the amplitude, a very sensitive feed mechanism, and an adjustable stop for the head.

Silicon carbide in aqueous suspension is used for making mosaics: 600 mesh for cutting slots and 1200 mesh for forming circular projections [50].

Ultrasonic Drilling of Diamond Dies and Cutting of Diamonds

Ultrasonic methods are suitable for drilling diamond dies, which are widely used in drawing fine wires of nonferrous metals. Ultrasonic methods can be used to drill new dies or to redrill old ones.

Ultrasonic methods give better shape and surface finish than do the mechanical and electrical methods, as well as providing higher cutting rates [51].

For instance, one ultrasonic machine in 25 shifts produced 25 diamond dies of diameter 0.6 mm, whereas it required 15 machines to produce the same number of dies in the same time by the mechanical method.

Ultrasonic machining has been applied to the lubricating and working cones, and also to the finishing section, in dies of diameter 0.3-1.2 mm.

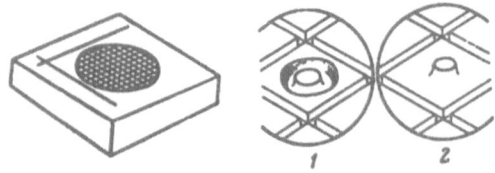

Fig. 151. Specimens of special shapes cut
from semiconductor plates.

* Areas 0.76 × 0.76 mm can be formed by the use of thinner blades and an abrasive of smaller grain size.

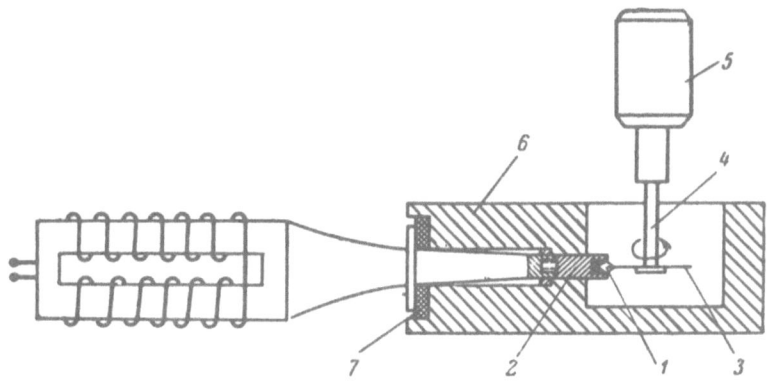

Fig. 152. Ultrasonic diamond slicing.

The tool is a steel needle sharpened to the required angle. High accuracy is ensured by rotating the die or tool at 100-110 rpm. The table must ensure that the die is self-centering to the needle. The feed force is 10-100 g, the die diameter being the decisive feature.

Diamond powder (grain size AM28) is used in water at 10-20% concentration. Up to 0.5 carat of diamond powder is consumed per die. The drilling of a new die of diameter 0.3-1.2 mm takes from 3 to 12 h, while re-machining of a worn one takes 1.5-3 h. The tolerance can be as low as 5-7 μ.

The die is polished in the usual way after ultrasonic machining.

Experiments on ultrasonic slicing of diamonds by the usual method (knife-type tool) gave a linear wear of 2000% on a tool 0.08 mm thick, which made the operation impracticable.

The knife tool was therefore replaced by a rotating thin steel disc, in which the wear is distributed over the entire perimeter and so is less important. A disc 30-45 mm in diameter is driven at 500-1000 rpm, which prevents the abrasive from settling out in the cutting vessel and also accelerates the process somewhat by purely mechanical cutting. The vibration (amplitude about 40 μ) is applied to the diamond. An aqueous suspension of A4 diamond powder (concentration 15-20%) was used, the feed force being 150-300 g. Powers of 0.25-0.4 kW suffice to cut stones of weight up to 2 carats; the cutting rate is then 0.2-0.33 mm^3/min. For instance, a 1.2 carat diamond was cut in 48 min with a disc 0.06 mm thick.

Figure 152 shows the apparatus used in these experiments on diamond cutting.

The diamond 1 is held by solder in the tip 2, which is screwed to the concentrator. The disc tool 3 is fixed to the shaft 4, which is driven by the motor 5. The vessel 6 holds the suspension and is attached via a flange to a node on the concentrator; the seal 7 prevents leakage of suspension. The feed is applied to the head via a weight. The crystal may be set relative to the tool by rotating the acoustic head and by moving the shaft vertically [52].

Machining of Ceramics

Ultrasonic machining is of value for the ceramic components used in electrical engineering and electronics.

For instance, the method has been used to generate slots 0.5 mm wide in ceramic inserts for mercury rectifiers and for porcelain coil formers for use with printed circuits.

Ultrasonic cutting has been used to cut parts from high-voltage porcelain insulators in order to determine the mechanical stresses. A circular groove of outside diameter 15 mm and width 0.5 mm was cut to a depth of 8 mm in 4-8 min with a 200 W machine, the tool wear being 4% [53].

Ultrasonic cutting is also applicable to cermet components consisting of the carbides, nitrides, silicides, and borides of the transition metals (tungsten, molybdenum, zirconium, titanium, niobium, and chromium). These materials are highly heat resisting and otherwise can be worked only by grinding.

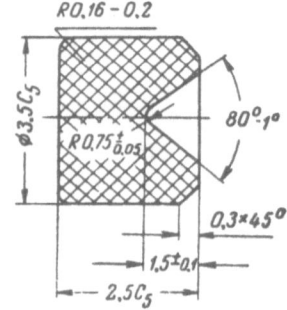

Fig. 153. Bearing of leuco-sapphire made by ultrasonic machining.

Tests have shown that the cutting rates with boron carbide as abrasive are 5-10 times those found for hard alloy. The cutting rate increases substantially with the porosity, and fracture occurs along the grain boundaries [54].

Metal-working tools can be sharpened by ultrasonic methods when these are tipped with corundum; chip-breakers can be formed in the same way. Mechanical sharpening involves the removal of a layer of material 0.2-0.8 mm thick on the front and rear faces.

Tests on this corundum (grade TsM332) have shown that ultrasonics gave a cutting rate 9-10 times that given by hard alloy type T15K6. For instance, a groove $14 \times 2.6 \times 0.2$ mm was cut in a plate in 6-7 sec, as against 50-60 sec for a hard-alloy plate.* The cutting rate on the corundum was 75-90 mm^3 per min, the tool wear being 0.7-2%, as against 100% on T15K6 alloy.

The height of the surface roughness at the cutting edge was 2-5 μ when boron carbide No. 6 was used, though scratches of depth 10-15 μ were also present. The surface finish is of class 7-8; microcracks, pits, and burrs were not seen.

Tools sharpened by ultrasonics have given satisfactory machining [55].

Ultrasonic Machining of Technical and Precious Stones

Technical stones (plain and thrust bearings) are much used in instruments; they are made from ruby, leuco-sapphire, and agate. These materials are cut and machined with diamond tools, which are cheaper than ultrasonic ones for long-run and mass production.

On the other hand, ultrasonic methods are of value for short-run production and for making components of complex shape. The uses of ultrasonics may be illustrated as follows.

Figure 153 shows a special sapphire bearing made in two steps; first the blank is cut, and then the conical hole is formed. Total machine time 1 min [4].

Watch jewels (blanks) of sapphire, agate, or ruby are made to tolerances of ±25 μ by ultrasonic methods. A tool with 35 holes in the end, each 1.6 mm in diameter, is used [56].

Holes 0.15-0.30 mm in diameter are drilled in rubies 0.5-0.8 mm thick without rotation of the tool in 2-3 min by means of boron carbide No. 3; the workpiece is vibrated, and the tool wear is 1.0-1.7 mm per hole.

Experiments on drilling even smaller holes (especially necessary in watchmaking) were not successful. Tools 0.05-0.08 mm in diameter wear very rapidly and often break.

Ultrasonic cutting of jasper, tourmaline, and sapphire increases productivity by factors of 5-8 and saves 30-40% in material; moreover, diamond is replaced by boron carbide. For instance, blanks of jasper and tourmaline 12 mm in diameter and 5 mm thick were cut in 2 min by a hollow tool with a wall thickness of 0.3 mm. Sapphire can be cut into plates 0.5 mm thick [57]. Engraving of gemstones has a rate of 0.025-0.5 mm/sec at a power of 200 W, the tool being good for 300 parts [58]. Moreover, the engraving is of very high quality. Jewelry of precious stones may be made with shaped holes to take gold parts [59].

Machining of Ferrites

Ultrasonic cutting and drilling have been used on ferrites for use in computers and various electronic devices (Fig. 154). These grooving and drilling operations are difficult to perform in any other way because ferrites are extremely brittle. A hole of diameter 0.38 ± 0.038 mm may be drilled to a depth of 2.3 mm in times of 30 sec to 6 min, the composition of the ferrite being the decisive factor. The tool is sometimes rotated during this operation [60].

---

* These experiments were done with a machine of power 0.6 kW at 20 kc with a feed force of 0.5-0.8 kg and boron carbide as abrasive (silicon carbide was also used, but this reduced the cutting rate by a factor of 1.6-1.8); the amplitude was 25-30 μ.

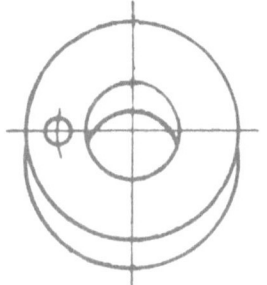

Fig. 154. Ferrite core with ultrasonically drilled holes.

Ultrasonic machining has also been applied to the profile cutting of cores for storage devices, in particular the cutting of grooves 0.038-0.040 mm wide [61].

It is difficult to ensure a high surface finish on a ferrite, on account of very ready cavitation erosion. This means that ultrasonic machining should be one of the earlier operations if high accuracy and good surface finish are essential.

Ferrites do not cut well in the ultrasonic method if they have any residual magnetization, because the cut particles are not readily removed from the cutting zone.

Ultrasonic Thread Generation

Section 18 deals with the kinematics of the forming of internal and external threads.

Male threads are generated by a tool taking the form of part of a die, while female threads are generated in much the usual way, except that the flutes for the chips in the shank are replaced by grooves to admit the suspension. The component must have already been drilled before a female thread can be generated. Thread-cutting tools are made of steel. Ultrasonic methods are not to be preferred for generating fine female threads, because the weak tools often break. The cutting times for threads 1/4 in. in diameter and 12.7 mm long in various materials are as follows:

Graphite	15 sec
Ferrite	16 sec
Glass	75 sec
Sintox	7 min
Hard alloy	15 min

Figure 155 shows a die made of hard alloy that was cut in this way [62]. Gumanyuk [63] has cut M4 threads on porcelain tubes by ultrasonic methods.

Machining of Steel Parts

Ultrasonic methods are occasionally applied to steel, although it does not cut freely.

For instance, multipoint steel dies for making small electronic plastic components have been finished in this way in Britain. A steel plate is first drilled with circular holes, after which the plate is hardened (R_c not less than 62) and is machined ultrasonically with a profile tool. The time to profile one hole about 4 mm in diameter to a depth of 3.16 mm is 8-9 min. A single cut gives the required size and taper. This method has proved time-saving and has eliminated hardening cracks [64].

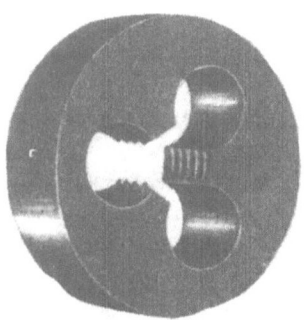

Fig. 155. Thread-cutting die made ultrasonically.

Ultrasonic methods have also been used to generate rectangular grooves in steel sleeves for the hydraulic control systems of jet aircraft and rockets. These grooves are 0.51-1.3 mm wide, up to 3.2 mm long, and 0.25-1.77 mm deep. Here two cuts are taken with boron carbide (280 and 800 mesh). The radius at the corners did not exceed 0.05 mm, the asymmetry of the slot relative to the axis did not exceed 0.005 mm, the walls deviated from perpendicular by not more than 0.025 mm, and pits and notches were absent [65, 66]. Recently the rough machining of these slots has been done by the electrospark method [67]. The ultrasonic machining made it possible to make these sleeves integral while providing good surface finish.

Ultrasonic methods are also very promising for the marking of small components, e.g., precision ball bearings. The tool is a replaceable steel needle.

Ultrasonic Drilling in Dentistry

Ultrasonic cutting is very promising here, because it enables the dentist to drill the tooth with a hole of any shape without producing pain. Further, the drilling does not generate heat, nor is there any danger of injury to soft tissues. Taking the cutting rate on cement as 100, the rate for glass becomes 35, that for dentine 23, enamel 21.5, amalgam 20, and autocrylate 10.4. The abrasive is electrocorundum. Various experimental ultrasonic drills have been made, and clinical studies are in progress [68].

Literature Cited

1. L. D. Rozenberg and D. F. Yakhimovich, "Current position and development prospects in ultrasonic cutting," in: Current Situation and Trends in Machine-Tool and Instrument Technology, Mashgiz, 1960, p. 260.

2. V. N. Barke and A. L. Livshits, "Current situation and trends in the ultrasonic machining of materials," in: Current Trends in Machine-Tool Technology, Mashgiz, 1957, p. 152.

3. E. A. Neppiras, "Report on ultrasonic machining," Metalwork. Product. 100(27-31, 33, 34):1283-1288, 1333-1336, 1377-1382, 1420-1424, 1464-1468, 1554-1560, 1599-1604, 1956.

4. L. D. Rozenberg and D. F. Yakhimovich, Ultrasonic Dimensional Machining of Brittle Materials, Profizdat, 1961.

5. A. Kuris, "New process for producing holes in hard metal," Machinery 57(10):175, 1951.

6. S. Levy, Method for removing material, US patent No. 2850854.

7. P. E. D'yachenko, Yu. N. Mizrokhi, and V. G. Aver'yanova, "Some aspects of the ultrasonic machining of materials," in: Use of Ultrasonics in Industry, Moscow, Mashgiz, 1959, p. 149.

8. V. V. Metelkin and I. V. Metelkin, "Tool for ultrasonic machining," Mashinostroitel' No. 5:35, 1958.

9. N. S. Goryachev, "Technique of machining hard-alloy press tools," in: Advances in Electrical and Ultrasonic Machining of Materials, Lenizdat, 1959, p. 183.

10. B. N. Mezhuev, "Ultrasonic machining of parts made of ceramic materials," in: Advances in Electrical and Ultrasonic Machining of Materials, Lenizdat, 1959, p. 203.

11. V. Yu. Veroman, Dimensional Ultrasonic Machining of Materials, Mashgiz, 1961.

12. L. Ya. Popilov, Electrical and Ultrasonic Machining: A Textbook, Moscow-Leningrad, Mashgiz, 1960.

13. Electrical-Pulse and Ultrasonic Methods of Machining Hard-Alloy Press Tools and Wire-Drawing Dies, ÉNIMS, 1960, Moscow.

14. E. A. Neppiras, Design of Ultrasonic Machine Tools, London, 1958.

15. V. Yu. Veroman, "Ultrasonic method of making hard-alloy dies," in: Advanced Scientific and Production Experience, No. M-60-29/2, Izd. TsITÉIN, 1960.

16. C. L. Calosi, Tool chuck for vibrating devices, US patent No. 2680333, dated March 16, 1951.

17. D. F. Yakhimovich, "Construction and design of the vibrating systems of acoustic heads for ultrasonic machine tools," in: Advanced Scientific and Production Experience, No. M-59-418/10, Izd. TsITÉIN, 1959.

18. I. V. Metelkin, V. E. Popov, V. I. Nikol'skii, V. V. Metelkin, and A. A. Mukaseev, "Mechanical machining of various materials with ultrasonics," Stanki i Instr. No. 2:16, 1956.

19. Technique of Ultrasonic Machining (Manuscripts) Moscow, TsBTI ÉNIMS, 1959.

20. C. Ballhausen, "Über die Anwendung von Ultraschall und Funken-Erosion zur Feinbearbeitung," Werkstattstechnik. u. Maschinenbau 45(11):557, 1955.

21. G. Pahlitzsch and D. Blanck, "Fortschritte beim Stossläppen mit Ultraschallfrequenz (Ultraschallbearbeitung)," Werkstattstechnik u. Maschinenbau 50(11):592, 1960.

22. "L'Usinage des métaux et corps durs sur la machine ultra-sonore Diatron," Ind. franc. Achats et entrét. matér. industr. 7(98):891, 895, 897, 899, 900, 1958.

23. V. Yu. Veroman, "Ultrasonic machining of hard alloys," in: Current Situation and Trends in Machine-Tool and Instrument Technology, Mashgiz, 1960, p. 190.

24. V. I. Kurchenko, Method of ultrasonic machining, Author's certificate, USSR No. 131210, dated September 7, 1959.

25. G. Nishimura and S. Shimakawa, "Ultrasonic mechanical machining. II. Surface finishing of hard metal by ultrasonic mechanical machining," J. Fac. Eng., Univ. Tokyo 25(1):47, 1957.

26. M. A. Bezborodov, A. A. Gezburg, and N. P. Krasnikov, "Use of ultrasound in the machining of glass," Steklo i Keram. No. 6:11, 1955.

27. A. L. Livshits, B. Kh. Mechetner, and V. N. Barke, "The 4772 universal ultrasonic machine tool," Stanki i Instr. No. 6:10, 1959.

28. P. Gagnaire, "Les ultrasons dans le cadre industriel," Ind. franc. Achats et entrét. matér. industr. 8(87): 507, 509, 511, 1959.

29. Methods of Accelerating Ultrasonic Machining (Manuscripts), Moscow, ÉNIMS, 1959.

30. G. Nishimura, K. Yanagishima, and T. Shima, Ultrasonic electro-spark machining, J. Fac. Eng.,Univ. Tokyo 25(1):41, 1957.

31. A. I. Markov and B. N. Lyamin, Method of ultrasonic machining, Author's certificate, USSR No. 109844, dated Jan. 14, 1957.

32. A. I. Markov, "A new method of increasing the cutting rate in ultrasonic machining," Vest. Mashinostr. No. 12:46, 1958.

33. A. I. Markov, V. I. Vasil'ev, and B. N. Lyamin, Mechanical Working of Hard Materials with Ultrasonic Vibrations, Moscow, No. B-57-30, Izd. Filiala VINITI, 1958.

34. Electrospark Working of Metals, Moscow, Izd. Akad. Nauk SSSR, 1960, issue 2.

35. N. S. Goryachev, Production of Hard-Alloy Dies by Ultrasonic Machining, Izd. Mosk. doma nauchno-tekhnicheskoi propagandy im. F. É. Dzerzhinskogo, 1957.

36. N. J. Clark and J. P. Aloisio, L'usinage par ultra-sons, Mach. Mod. No. 550:7, 1955.

37. W. Lehfeldt, Die Bedeutung des Ultraschalls in der Feinbearbeitung," Techn. Mitt ,50(1):28, 1957.

38. E. J. Tangerman,"Work cuts tool, tool forms work – with ultrasonic," Machinist 98(52):2275, 1954.

39. W. M. Stocker, Ultrasonic cutting: relief for hard-material headaches," Metalwork. Product. 99(31): 1379, 1955.

40. Yu. M. Anserov and É. G. Ter-Zakharyan, "Ultrasonic machining of brittle nonmetallic materials," Mashinostroitel' No. 5:33, 1959.

41. G. Nishimura and S. Shimakawa, "Ultrasonic Mechanical Machining. V. Circular Trueness of Glass Hole Drilled by Ultrasonic Mechanical Machining, " J. Fac. Eng.,Univ. Tokyo 26(1):1, 1959.

42. V. V. Kupfer, "High-production system for engraving glass components," in: Current Situation and Trends in Machine-Tool and Instrument Technology, Mashgiz, 1960, p. 223.

43. M. M. Pisarevskii and A. A. Klenov, "Experience with the drilling of holes in quartz plates by ultrasonics," in: Use of Ultrasonics for Machining Hard Materials, Izd. Mosk. doma nauchno-tekhnicheskoi propagandy im. F. É. Dzerzhinskogo, 1957, p. 2.

44. E. N. Gibbs, "Ultrasonic cutting of quartz wafers," Ultrasonic Eng. No. 4:66, 75, 1956.

45. "Ultrasonic slicer cuts quartz crystals," Tele-Tech & Electronic Ind. 15(7):15, 1956.

46. M. S. Hartley, "Ultrasonic machining of brittle materials," Electronics 29(1):132, 1956.

47. "Electromechanical methods for machining and grinding cemented carbides," Machinery (London) 83 (2143):1146, 1953.

48. "Ultraschall bei der Germanium-Bearbeitung," Funkschau, 29(2):34, 1957.

49. "Stainless-steel tubing used for ultrasonic machining tools," Machinery (USA) 63(1):161, 1956.

50. K. D. Knight, "Production of islands and dice in semi-conductor slices with an ultrasonic drill," Sci. Instr. 37:263, 1960.

51. I. V. Stroganov, "Ultrasonic drilling and polishing of diamonds and hard alloys," Kabel'n. tekhn. No. 2(9): 49, 1959.

52. P. E. D'yachenko and Yu. N. Mizrokhi, "Machining of diamonds with ultrasonic vibrations and a rotating disc," Vest. Mashinostr. No. 3:60,1960.

53. V. N. Shchepetov, "Ultrasonic apparatus for cutting ceramic materials," in: Advanced Scientific and Production Experience, Moscow, No. M-57-126/1, Izd. Filiala VINITI, 1957.

54. A. A. Mukaseev, V. S. Rakovskii, B. N. Babich, and Yu. V. Levitskii, "Some aspects of the machining of refractory cermet materials with ultrasonics," Vest. Mashinostr. No. 3:67, 1961.

55. V. N. Berezub, A. E. Potapenko, and E. S. Chistyakov, "Ultrasonic method of sharpening a cutting instrument," Vest. Mashinostr. No. 3:67, 1961.

56. R. Singer, "Les faconnages pour hautes fréquences ultra-sonores," Electricien 87 (1987) : 42, (1959).

57. A. P. Sviridov, "Semiautomatic ultrasonic machine," Izv. Vyssh. Uch. Zav. Priborostroenie No. 2:159, 1958.

58. J. Welch, "Des utilisations toujours nouvelles de l'usinage par ultra-sons," Mach. Mod. 52(591):9, 1958.
59. Brazing and Soldering, Am. Machinist 97(24):E4-E5, 1953.
60. N. K. Marshall, "Drilling small holes by the ultrasonic method," Machinery(Engl.)32(2361):379, 1958.
61. "Ultrasonic machining of ferrite," Electron. Design 5(22):94, 1957.
62. "Thread Generation by Ultrasonics," Machinery Lloyd, Overseas Ed. 30(11):90, 1958.
63. M. N. Gumanyuk, "Use of ultrasonics in technological processes," Byul. Tekhn.-Ékon. Inform. Sov. Nar. Khoz. Khar'kovskogo Ékon. Admin. Raiona, No. 1:26, 1958.
64. R. V. G. Elwes, "Ultrasonic machining," Plastics 22(236), 1957.
65. W. W. Wood, "Ultrasonic machine cuts precision holes in hydraulic sleeves," Machinery (USA), 61(10): 220, 1955.
66. A. E. Shumate, "Ultrasonic impact grinding produces precise shaped holes, slots on servo-valves with ease," Western Metals 15(2):41, 1957.
67. "Electrical discharge and ultrasonics form hard-working team, Machinery (USA) 64(9):195, 1958.
68. L. Balamuth, "Technical aspects of the cavitron ultrasonic process in dentistry," IRE Convent. Rec. March 21-24:89, 1955.
69. I. M. Gissin, "Shape of the edge of the workpiece in ultrasonic grinding," in: Ultrasonic and Electroerosion Methods of Machining Metals, Rostov-on-Don, 1961, pp. 111-118.
70. A. I. Markov, Cutting of Difficult Materials by Means of Ultrasonic Vibrations, Mashgiz, Moscow, 1962.